ROUTLEDGE LIBRARY EDITIONS:
HUMAN GEOGRAPHY

Volume 6

BEHAVIORAL PROBLEMS IN GEOGRAPHY REVISITED

BEHAVIORAL PROBLEMS IN GEOGRAPHY REVISITED

Edited by
KEVIN R. COX AND
REGINALD G. GOLLEDGE

Routledge
Taylor & Francis Group

LONDON AND NEW YORK

First published in 1981 by Methuen & Co. Ltd

This edition first published in 2016
by Routledge
2 Park Square, Milton Park, Abingdon, Oxon OX14 4RN

and by Routledge
711 Third Avenue, New York, NY 10017

Routledge is an imprint of the Taylor & Francis Group, an informa business

British Library Cataloguing in Publication Data
A catalogue record for this book is available from the British Library

ISBN: 978-1-138-95340-6 (Set)
ISBN: 978-1-315-65887-2 (Set) (ebk)
ISBN: 978-1-138-95126-6 (Volume 6) (hbk)
ISBN: 978-1-315-66831-4 (Volume 6) (ebk)

Publisher's Note
The publisher has gone to great lengths to ensure the quality of this reprint but
points out that some imperfections in the original copies may be apparent.

Disclaimer
The publisher has made every effort to trace copyright holders and would welcome
correspondence from those they have been unable to trace.

Behavioral Problems in Geography Revisited

edited by
KEVIN R. COX
and
REGINALD G. GOLLEDGE

METHUEN

NEW YORK AND LONDON

First published in 1981 by
Methuen & Co. Ltd
11 New Fetter Lane, London EC4P 4EE

Published in the USA by
Methuen & Co.
in association with Methuen, Inc.
733 Third Avenue, New York, NY 10017

Printed in Great Britain by
Richard Clay (The Chaucer Press) Ltd
Bungay, Suffolk

British Library Cataloguing in Publication Data

Behavioral problems in geography revisited.
1. Anthropo-geography – Addresses, essays, lectures
I. Cox, Kevin R.   II. Golledge, Reginald G.
909 GF41

ISBN 0-416-72430-2
ISBN 0-416-72440-X Pbk (University paperback 685)

Contents

List of illustrations vii
Notes on contributors ix
Preface xiii

PART 1

1 Inference problems in locational analysis 3
Gunnar Olsson

2 Conceptual and measurement problems in the
cognitive–behavioral approach to location theory 18
David Harvey

3 The geographical relevance of some learning theories 43
Reginald G. Golledge

4 The scaling of locational preferences 67
Gerard Rushton

PART 2

5 Cognitive mapping: a thematic analysis 95
Roger M. Downs

6 Behavioral approaches to the geographic study of
innovation diffusion: problems and prospects 123
Marilyn A. Brown

7 Cognitive behavioral geography and repetitive travel 145
John S. Pipkin

8   Residential mobility and behavioral geography:
    parallelism or independence?   182
    W. A. V. Clark

    PART 3

9   Behavioral geography and the philosophies of meaning 209
    David Ley

10 Of paths and projects: individual behavior and its
    societal context   231
    Allan Pred

11 Bourgeois thought and the behavioral geography debate 256
    Kevin R. Cox

    Author index   281

    Subject index   288

List of illustrations

TABLES

3.1 Cues for selection of shopping centers 55
3.2 Paired comparison of shopping centers 58
4.1 Revealed space preference, raw data matrix by respondents 74
4.2 Segments of the revealed preference data matrix 76
4.3 Probability that column locational type is preferred to row type 78
4.4 Scale values for the locational types 82
4.5 Transitivity test for consistency of preference surface 85
4.6 A response matrix showing responses to various locational types by various individuals on a single occasion 87
4.7 A response matrix showing responses to various locational types on various occasions by the same individual 88

FIGURES

4.1 Definition of locational types 71
4.2 Space preference structure for grocery purchases: Iowa, 1960 83
6.1 A portrayal of the interface between diffusion agency actions and adoption behavior 130

10.1 The activity bundles and path convergences associated with a hypothetical organizationally defined project stretching over several days 238

10.2 An overview of individual behavior and its societal context based on time-geography and *genre de vie* concepts 248

10.3 Törnqvist's 'observation field' overview of the 'registered behavior' of individuals and its societal context 250

Notes on contributors

Marilyn A. Brown is currently Assistant Professor of Geography at the University of Illinois at Urbana-Champaign. She obtained her doctorate from the Ohio State University. Her research interests are in social psychological models of the adoption and non-adoption of innovations, cognitive mapping, and contemporary urban and suburban morphology. She has published articles on each of these topics in a variety of journals.

W. A. V. Clark is Professor of Geography at the University of California, Los Angeles. He taught formerly at the Universities of Canterbury and Auckland in New Zealand, at the University of Wisconsin, Madison, and at the Free University of Amsterdam in the Netherlands. His current interests are focused on intraurban migration and neighborhood change. He is co-author of *Los Angeles: The Metropolitan Experience* (Cambridge, Mass., Ballinger, 1976) and co-editor of *Population Mobility and Residential Change* (Chicago, Northwestern University, 1978) and *Residential Mobility and Public Policy* (Beverly Hills, Ca., Sage, 1980).

Kevin R. Cox is Professor of Geography at the Ohio State University. His primary research interests are urban conflict and urban social thought. His papers have appeared in professional journals in geography, sociology and political science. He is co-editor of *Behavioural Problems in Geography: A Symposium* (Chicago, Northwestern University, 1969), co-editor of *Locational Approaches to Power and Conflict* (Beverly Hills, Ca., Sage,

1974), editor of *Urbanization and Conflict in Market Societies* (London, Methuen, 1978) and author of *Conflict, Power and Politics in the City: A Geographic View* (New York, McGraw-Hill, 1973).

Roger M. Downs is Professor of Geography at the Pennsylvania State University and has taught at Johns Hopkins University, Colgate University and the University of Washington. His current research interests include the relationships between cognition and cartography, the concept of cognitive mapping and the design of human wayfinding systems. Together with David Stea, he has published *Image and Environment* (Chicago, Aldine, 1973) and *Maps in Minds* (New York, Harper & Row, 1977).

Reginald G. Golledge is Professor of Geography at the University of California at Santa Barbara. His main research interests are in urban geography, behavioral geography and cognitive mapping. His current work involves determining the spatial competence of socioeconomically and educationally deprived subgroups and the borderline mentally retarded.

David Harvey is Professor of Geography and Environmental Engineering at Johns Hopkins University. Previously he taught at Bristol University and at Pennsylvania State University. His papers have appeared in geography, political science and regional science journals. He is the author of *Explanation in Geography* (London, Arnold, 1969) and *Social Justice in the City* (London, Arnold, 1973).

David Ley is Associate Professor of Geography at the University of British Columbia in Vancouver, Canada. His research interests are in urban and social geography, particularly inner-city issues, and in social theory and geography. He is the author of *The Black Inner City as Frontier Outpost* (Washington, D.C., AAG, 1974) and *A Social Geography of the City* (forthcoming, 1982), editor of *Community Participation and the Spatial Order of the City* and joint editor of *Humanistic Geography* (Chicago, Maaroufa, 1978).

Gunnar Olsson is currently Professor of Economic Geography and Planning at the Nordic Institute for Planning in Stockholm, Sweden. He previously taught for eleven years at the University of Michigan. His publications include *Distance and Human Interaction*

(Philadelphia, Pa., Regional Science Research Institute, 1965), *Philosophy in Geography* (co-edited with S. Gale) (Dordrecht, Reidel, 1979), *Birds in Egg/Eggs in Bird* (London, Pion, 1980) and *Search for Common Ground* (co-edited with P. Gould).

John S. Pipkin is with the Geography Department in the State University of New York at Albany. His interests include aggregate and disaggregate spatial choice, space cognition, and urban retail structure. In addition to articles in these areas, he co-authored *Urban Social Space* (Belmont, Ca., Wadsworth, 1981) and is co-editing a forthcoming collection on urban planning.

Allan Pred is Professor and Chairman in the Department of Geography, University of California, Berkeley. He has published extensively on the growth and development of city systems and on other time–space processes. His latest book is *Urban Growth and City Systems in the United States, 1840–60* (Cambridge, Mass., Harvard University Press, 1980). His current research and writing is aimed at an integration of time geography and social theory.

Gerard Rushton is Professor of Geography at the University of Iowa, Iowa City, Iowa. A co-author of *Spatial Choice and Spatial Behavior* (Columbus, Ohio State University Press, 1978), he is currently researching the determinants of migration of the elderly.

Preface

The specific point of departure for this collection of essays is a volume brought together by the present editors and published in 1969 under the title *Behavioral Problems in Geography: A Symposium*. The book both expressed and, we think we can fairly claim, gave impetus to the then powerful tide of interest in geographic theory and, more precisely, the elaboration of adequate behavioral postulates for such theory. The collection was widely referenced and received favorable reviews.[1]

Just over ten years have elapsed since the appearance of that book. During this interval the field of geography has developed in ways that few could have foreseen. In particular, while the 1960s can be justifiably labeled the period in which human geography became a theoretically self-aware discipline, the 1970s subverted our new-found confidence with periodic fits of self-consciousness at the epistemological level: geography as a science was followed by self-doubt as to what sort of science geography could and should be.

It would be surprising if these shifts within the discipline had not left their mark upon behavioral geography.[2] Indeed, its self-confidence and brash exuberance have been major casualties of these challenges. Symbolic of this are the intellectual trajectories of two of those geographers who were most closely associated with behavioral geography: David Harvey and Gunnar Olsson. Both wrote papers for the 1969 volume. Their papers were widely referenced and, indeed, are reprinted in this collection. Today, neither would acknowledge the label 'behavioral geographer':[3] in the past decade both have shifted from the role of prophet to that of critic. In a real sense, their shift epitomizes the epistemological challenge which behavioral geography must now confront.

Yet behavioral geography has not only survived; it has also developed. The mere trickle of papers existing in the late 1960s on environmental cognition turned into a veritable flood. Papers written at that time had *major* impacts upon work in fields to which one readily turns for examples of work in behavioral geography: the impacts of Hägerstrand upon spatial diffusion research[4] and of Wolpert upon migration[5] are exemplary in this regard.

A re-evaluation, therefore, seems appropriate, and this is the aim of the current book. Specifically what the book does is three things. Firstly, it provides some sense of what behavioral geography was about as it developed in the mid to late sixties. This is done by reprinting, completely or in part, four of the most widely referenced papers from the 1969 volume: the papers by Olsson, Harvey, Golledge and Rushton. These comprise Part 1 of the present book.

Secondly, we have sought to provide some sense of the impact of behavioral work in four major areas of endeavor: intraurban mobility, spatial diffusion, environmental cognition and urban travel behavior. Toward this end Part 2 of the book includes papers by W. A. V. Clark, Marilyn Brown, Roger Downs and John Pipkin. Finally, and in recognition of the serious epistemological challenges of the seventies, Part 3 provides papers of a more critical character. Thus David Ley and Kevin Cox discuss behavioral geography from the standpoints of humanistic geography and dialectical materialism respectively, while Allan Pred treats a critique more internal to behavioral geography–time geography.

In brief, Part 1 is of largely historical interest. Part 2 should appeal to those interested in specific fields of application in behavioral geography and also to those concerned with finding out how behavioral geography has impacted in those specific areas. Part 3 should be seen as complementary to the other parts in the sense that it provides a set of critiques of the work exemplified therein. First, in the remainder of this Preface, however, it seems important to provide an overview. In the initial section, therefore, we try to sketch out behavioral geography as it appeared ten years ago. In the second section we provide a brief summary and interpretation of what has happened since. In the final section we look towards the future in the light of the challenges behavioral geography now faces, both from within and from without.

Behavioral geography: the original conception

The papers brought together in the 1969 volume were all representative of a growing interest on the part of geographers in what was then loosely termed the behavioral approach to geographical problems. Eight of the papers had been originally presented at an arranged session at the annual meetings of the Association of American Geographers, Washington, D.C., 1968. The two additional papers – those of Harvey and Olsson – were solicited for the meeting but were not available at that time.

Contributions to the volume varied a great deal in their emphasis, ranging from philosophy and review to theorizing and operationalization. Each paper, however, recognized the importance of examining the behavioral basis of spatial activity. This was not new in and of itself, for many existing theories and models in geography had, and continue to have, at least an implicit behavioral element in their structure. What was new was the deliberate attempt to unpack and identify these elements, to examine their specific effects on spatial activity, and to operationalize some of the concepts previously used in a subjective and descriptive manner. All papers, therefore, were united by a common concern for the building of geographic theory on the basis of postulates regarding human behavior.

Consider first the role of behavioral models as it was seen at that time. Papers by Harvey,[6] and Olsson and Gale,[7] and a monograph by Pred[8] had all attempted to answer the questions of need and relevance. For example, as Harvey had stated, the postulates of geographic theory are in part indigenous and in part derivative: 'Indigenous geographic theory may be regarded . . . as an attempt to state the laws of spatial form in the specialized languages of geometry or topology, or in the more general form of spatial statistics.' He also pointed out, however, that such laws tell one little or nothing about processes, and for this geographic theory must turn to the other social sciences and the postulates about human behavior that they can provide. In order to understand spatial structure, therefore, one had to know something of the antecedent decisions and behaviors which arrange phenomena over space. This was the viewpoint stressed in the 1969 volume.

Thus, for example, it was clear that a large number of behavioral mechanisms had spatial correlates. This was particularly evident, for instance, in those studies making use of the concept of information flow where the sender and receiver of the information could be identified as to location and where the efficacy of the mechanism could be related to relative location:[9]

There was also at that time a growing number of studies in geography focusing upon the mental storage of spatial information.[10] Mental-map studies and action-space studies, for example, had and have implicit within them some relationship to the movements and activities that produce spatial structure, and spatial structure could not be understood without some knowledge of the perception of spatial reality retained in the human mind. Hence the emphasis in the 1969 volume was upon social and psychological mechanisms having explicit spatial correlates and/or spatial structural implications. In this sense the work differed from some of the work done in geography upon environmental perception, where the interest was less in behavioral mechanisms having spatial correlates and more upon the measurement of attitudes toward environmental stimuli.[11]

An important distinction emergent in behavioral geography at that time, and one underlining the distinctive ethos of behavioral work, was that between studies of spatial behavior and of behavior in space. The basic level of research was provided by studies of spatial behavior. This level consisted of a search for relevant postulates and models to describe behavioral processes irrespective of the spatial structure in which the behaviors were found. In other words this search was for the rules of choice, movement and interaction that are independent of the spatial system in which they are to operate. Such rules, rather than descriptive statistics of actual behavior in a system, should, it was felt, provide the basis for future theorizing. Rushton's paper from the 1969 volume, which provided a first outline of a method for identifying such rules, is reprinted in this volume.

A second level related parameters describing actual behavior in an area to specified spatial structures in the same area, i.e. studies of behavior in space. An excellent example of this level could be seen in Hägerstrand's use of mean information fields, where the parameters of the field were based upon actual interaction data for the spatial system in question.[12] Such parameters, as Rushton suggested,[13] were most likely to be related to the spatial structure of interaction opportunities in the specific spatial system. Under such circumstances behavior in space was constrained by the existing spatial structure and should not be used to explain that structure.

It is the idea of spatial behavior rather than behavior in space, therefore, that provided the inspiration for behavioral geography. It is, moreover, an inspiration that has found reflection elsewhere in geography among those who would not regard themselves as behavioral geographers. The debate over the interpretation of gravity-model coefficients is a significant case in point[14] and underlines the centrality of the themes and conceptual distinctions identified by behavioral work to the evolution of human geography over the last decade.

In retrospect, there turned out to be a second important distinction. This was the emphasis upon the individual as active rather than reactive. This gave behavioral geography in North America a particular stamp which was less apparent elsewhere – in Britain and Australia, for example. It is a stamp that is clearly apparent in this volume.

Thus in a survey of behavioral geography Thrift[15] notes that behavioral research in the United Kingdom has become neither as widespread nor as commonly accepted as its equivalent in North America. One of the reasons given for this includes a more limited view of behavioral research in the UK: in particular, a tendency to place an emphasis on the individual as a *reactive* rather than *active* decision maker. The North American focus has been on individuals as *active* decision makers processing information through different perceptual filters and choosing between alternatives. Decision making, choice theory, analysis of space preference, spatial learning processes, environmental cognition and cognitive mapping have all developed out of the process of identifying the active decision-making component of man's behavior.

An alternative view that, Thrift argues, has been more characteristic of behavioral research in the United Kingdom assumes a model of man as a *reactive* decision maker subject to a number of socio-spatial constraints. This tradition had its roots in the 'activity analyses' of Chapin and his co-workers.[16] Critical in establishing it as a major research area was Hägerstrand's presidential address to the Regional Science Association in 1970.[17]

The reactive approach is more concerned with tracking individual behavior as an object of study in itself, and explaining man's actions more in terms of society. Consequently, whereas much of the *active* research focused on the individual, this emphasis was neither as necessary nor as frequently adopted by reactive researchers. Whereas the first group (perforce) required considerable attention to be paid to appropriate selection of methodologies suited to collecting and analyzing individual behavior, and was obsessed with problems of grouping or aggregating their individualized data sets, the *reactive* researcher was more comfortable starting from a group base and consequently focusing more attention on the societal, institutional and environmental constraints that limit actual and possible behaviors. Some sense of the degree to which this type of approach is regarded as outside the mainstream of behavioral geography in North America is that the only representative of it in this volume (Pred's paper) appears in the more critical Part 3.[18]

Behavioral geography: an evaluation

Consider now the accomplishments of behavioral geography. Three major achievements can be identified. The first is a general sensitivity of human geographers to the importance of theory. The procedure (classic in the early 1960s) whereby certain hypotheses – often isolated from and logically inconsistent with each other – were defined and tested has given way to a realization that testing and identification and interpretation of parameter values must be deferred until adequate theory is formulated. There has, consequently, been a sensitizing to the necessity for conceptualization and theoretical elaboration. The early work of Brown and Moore[19] and of Wolpert[20] is exemplary in this regard.

The second accomplishment was to underline the artificiality of the boundaries between the present systematic fields of geography. Thus, the learning models stressed by Golledge in the paper reprinted in this volume as applicable to shopping-center patronage and agricultural marketing decisions are also applicable to agricultural land-use decisions,[21] and indeed to any spatial diffusion problem.[22] The mental-map literature has implications for migration and also for an understanding of urban travel movements,[23] while the notion of information flow is relevant not only to the spatial diffusion of innovations[24] and the study of voting-response surfaces,[25] but also to economic geography.[26]

Thirdly and finally, there has been a more general sensitizing to the significance of behavioral postulates. This is nowhere more apparent than in the approaches of the diverse introductory human geography texts that began to appear in the early 1970s.

Yet one of the goals expressed in our 1969 editorial introduction has proven elusive: that of producing new and more powerful theories to explain human spatial behavior. This issue deserves a more lengthy treatment. We would argue, however, that a failure to contribute to the production of new theory need not necessarily be regarded as a major criticism. This is only a decade later, and if we examine the general literature of geography we would be hard put to define powerful new theory in any area of the discipline. This lack, therefore, might not be a symptom of an ailing patient but an indication of a general malaise that has had the discipline in its grip during these last ten years. Perhaps we should at this stage speculate a little on why the behavioral approach might not have lived up to all its expectations. In speculating we concentrate on a major area to which the behavioral approach hoped to contribute significantly: cognitive mapping.

The concept of a 'mental map,' while raised effectively by Tolman[27] in the psychology literature in 1948, had no geographic counterpart until a decade ago. Whether or not cognitive images were map-like in form, whether or not people could externalize cognitive information and schemata in map-like forms, and whether this was potentially of interest to the geographer, was almost completely unknown. It has taken a number of individual, isolated and repeated studies to get to the point where some conclusions can be drawn about the validity and usefulness of external representations of cognitive information. This phase has been reached only after extensive experimentation. This experimentation has been piecemeal and has had to start from fundamental levels. For some time even the definition of the term 'mental map' was confused, and the discipline at large only became acquainted with the nature of the difference between cognitive mapping and statements of preferences for places after the publication of *Image and Environment*.[28]

Apart from confusion raised by definitional problems, however, the geographer had to deal with a variety of concepts that existed in psychology but which had no acknowledged existence in the world of the geographer. These had to be identified, their relevance confirmed, data collected about them and the essence of their meaningfulness extracted. All this was necessary before an assessment could be made as to whether or not the concepts were useful in the realm of geographic explanation. The behaviorally oriented researcher, then, had to start from zero: he had little to build on, for the relevance of spatial concepts was relatively uninvestigated in the field of psychology. Within geography, the approach served to raise fundamental and still-unresolved questions, such as: 'What is distance?' 'What does proximity imply?' 'How can patterns be compared for similarity?' 'What is the spatial interpretation of utility?' 'Is there a spatial component to learning activities?' 'How does environmental information accumulate over time?' 'From the information that is received by an individual, how can we spatially represent what is known?' 'What is the philosophical distinction between real and perceived environments?' There were, in fact, many fundamental, definitional, theoretical, philosophical, epistemological, analytical and empirical questions to be solved before any of the major goals of the behavioral approach could be achieved. One should perhaps be surprised, therefore, that those working in the field have been able to contribute as much to the general realm of geographic understanding over the last decade as they have.

The future of behavioral geography: evolution or irrelevance?

Much of geography in the 1960s was dominated by an instrumentalist philosophy. While there was an increasing recognition that human spatial phenomena were not likely to be subject to universal laws in the same sense that physical spatial phenomena might be, researchers both sought for laws that might possibly exist and regarded human spatial phenomena *as if* they were subject to sets of universal laws. This produced a series of positions concerning the importance of spatial processes which for the most part were nonhuman and system oriented. Even when an emphasis was placed on process, it was in the context of attempting to define a particular spatial form or pattern. In many cases, it was also tied to a desire to produce a predictive or forecasting ability. In Olsson's terms,[29] the 1960s were dominated by the geographer as 'social engineer' whose closest contacts were with other like-minded disciplines such as regional science, and economics.

An important aspect of the intellectual environment within which behavioral geography situated itself in the late 1960s was precisely this instrumentalist view. Equally important was a concept of behavior and decision making intrinsic to the location theory which geographers adopted as the explanatory axis of locational analysis. This location theory was in turn the attempt to project on to space the assumptions of neoclassical economics. According to this, locations could be explained as the outcome of an individualistic, decentralized decision-making process in which perfectly informed, utility-maximizing, economic men made locational adjustments in response to spatial disequilibria in market prices; these locational adjustments in turn re-established equilibrium. Clearly this involved certain behavioral assumptions such as perfect knowledge and an ability to calculate optimal locations. Behavioral geography emerged from the recognition that these assumptions as comments upon real-world relationships left a great deal to be desired.[30] As such they would severely qualify the predictive power of location theory. In this sense, as Cox remarks in his paper in this book, behavioral geography was far from revolutionary.

Two characteristics of the intellectual milieu in which behavioral geography was situated imply alternative futures. The first is that behavioral geography could be voided as a viable intellectual project by a successful challenge to its epistemological assumptions, i.e. to those assumptions regarding subject–object separation, separation of individual from society, and of fact from value upon which the naturalism and instrumentalism of locational analysis (and hence behavioral geography) have been established. The second, however, is that this epistemological challenge could be successfully repulsed. Change within behavioral geography would then occur as a result of a critique or critiques within its own epistemological terms of reference. This might occur through an expansion of the list of relevant variables, for example, or a switch from an emphasis upon the individual as an isolated decision maker to the individual as caught up in a web of social constraint. Consider here, therefore, the alternative possibilities of evolution through internal critique, and of irrelevance through external or epistemological critique.

Evolution

Some early presentiments of the limitations of behavioral geography as it was emerging from location theory were in fact made by one of the early pioneers in the field: Julian Wolpert. Starting from the observation that location processes were often conflict ridden, he went on to indicate that the subsequent location pattern was often less the product of competition between individuals and more the outcome of bargaining between groups. As he remarked in a paper in the original *Behavioral Problems in Geography*:

> In the early stages of location research, the emphasis has been on normative models of rational decision making based on the twin assumptions of utility maximization and the independence of the participants in the decisional process. Subsequent work, using a behavioral approach, has modified this model by taking account of the multiple goals and restricted capacity of decision makers which limit the applicability of maximization or least cost solutions as these are usually understood. We now suggest that more serious limitations of the normative model stem from the fact that the participants in locational decisions are often groups whose relative organization, sophistication and power *develop* through the decision process, and whose interests are often conflicting but *interdependent*. In such game-like situations, which can be broadly conceived as bargaining structures, the rationality assumptions of classical economic theory break down.[31]

Underlying this idea seems to be a view of location processes much truer to the pluralistic and politicized circumstances of the time than the almost nineteenth-century character of location theory. This is a view, however, that has never clearly been articulated and developed within geography.

Hägerstrand's post-behavioral trajectory is in some ways very similar to that of Wolpert. Like Wolpert, Hägerstrand can be regarded as one of the founding fathers of behavioral geography. Likewise, Hägerstrand also realized, round about the turn of the decade, that an explanatory framework reliant upon a notion of decentralized, individualistic decision making had limited relevance to location in the late twentieth century: in some way the structuring effects of social relations had to be given greater prominence in geographic theory.

Hägerstrand's answer to this has been what has come to be known as time geography. Some sense of Hägerstrand's attempt to anchor his idea in a conception of social constraint is provided by the following quote from one of his earlier programmatic statements:

> When Robinson Crusoe found himself alone on his island, he could make up his program without regard to a pre-existing socio-economic system. The natural resources were all his to develop under his specific set of biological and technical constraints. An individual who migrates into an established society, either by being born into it or by moving into it from outside, is in a very different position. He will at once find that the set of potentially possible actions is severely restricted by the presence of other people and by a maze of cultural and legal rules. In this way, the life paths become captured within a net of constraints, some of which are imposed by physiological and physical necessities and some imposed by private and common decisions. Constraints can become imposed by society and interact against the will of the individual. An individual can never free himself from such constraints. [32]

In order to investigate the channeling effects of 'general and abstract rules of behavior' on individual location, Hägerstrand proposed to give them 'a "physical" shape in terms of space, areal extensions and duration in time.' [33] Thus, 'authority constraints' received a physical expression in the form of domains or time–space entities 'within which things and events are under the control of a given individual or a given group;' [34] 'coupling constraints' were represented by bundles of complementary roles and 'define where, when and for how long the individual has to join other individuals, tools and materials in order to produce, consume and transact.' [35]

The work of Wolpert and Hägerstrand has, in our judgement, had a modest impact upon the work of those in North America starting from an initial theoretical basis in behavioral geography. The more theoretically articulated notions of Hägerstrand have been taken up by one of the pioneers of behavioral geography, Allan Pred, [36] and he has contributed a paper along those lines to the present book. Likewise, as the papers by Brown (Chapter 6) and Clark (Chapter 8) in this book indicate, the early formulations of Wolpert persist in the attempt to qualify the individualistic character of behavioral explanation by reference to so-called 'institutional' factors. Nevertheless, in the light of the oft-heard pleas from behavioral geographers for greater attention to the 'social' dimension,[37] the impact has been less than might have been expected.

This is not to argue, however, that in their attempt to qualify the individualistic, decentralized decision-making emphasis of behavioral geography Wolpert and Hägerstrand were reacting to dissatisfactions of a purely intellectual character. Rather, ideas were making contact with the real world and, in the late 1960s, being found wanting. Hägerstrand confronted the issues directly as a result of his involvement in planning in Sweden. As Pred has written:

> Hägerstrand's call for a time-geographic focus on people and, in particular, the event-sequences which constitute the days and life of each individual person stems from a humanistic concern with the 'quality of life' and everyday freedom of action implications *for individuals* of both existing and alternative technologies, institutions, organizations and urban forms.[38]

Likewise, it is clear from Wolpert's own writings that in formulating his ideas he was highly influenced by the apparent contradiction between the predictions of location theory and the events emerging in the American city in the 1960s and early 1970s.[39] In particular, he was impressed by the conflict-ridden nature of public-facility location and its distributional implications. Also impressed was Harvey, who, in his liberal phase,[40] made points similar to those Wolpert had been making. This, however, brings us to the epistemological challenges that behavioral geography must confront.

Epistemological challenge

In contrast to Wolpert, Harvey went on to provide a critique of location theory external, rather than internal, to it. Epistemologically, Harvey proceeded to throw out the subject–object separation and fact–value distinction upon which location theory and locational analysis had been predicated, and which had been preserved by Wolpert and Hägerstrand, and to erect in its place a theoretical structure of marxian provenance.[41] This represented the first challenge to the epistemological hegemony of positivism and hence to behavioral geography. For Harvey was able to show, to the satisfaction of some geographers at least, that the individual was always a *social* individual. Consequently the rigid separation of individual from society that underlay both behavioral geography and the respective projects of Wolpert and Hägerstrand was epistemologically invalid; and, moreover, told us more about the way the world *appeared* than about the deep structures making those appearances both possible and meaningful. The phenomenal forms of decentralized, atomized decision makers that had made location theory and behavioral geography so plausible were contradicted by an underlying unity of individual and society that made itself periodically apparent in the form of crisis and upheaval. Accordingly, subject and object were mutually presupposing and not independent from one another as assumed by naturalistic and instrumentalist philosophies. The marxist critique of behavioral geography is developed at greater length in this volume in the essay by Cox (Chapter 10).

The second epistemological challenge comes from humanistic geography. Like marxism, humanism takes as its epistemological *bêtes noires* the subject–object and fact–value dichotomies. As Entrikin has written,

> Humanist geographers hold that the study of human behavior cannot be modeled after the physical sciences. They reject the positivist claim of the isomorphism between physical and social science, because they are dissatisfied with, among other things, two related dichotomies: the subject–object distinction and the fact–value distinction. These distinctions are related in that, by viewing the world as separable into the objective world of things and the subjective world of the mind, one can then separate the knowledge of that objective world as factual knowledge and the subjective element as emotion, value and meaning.[42]

Central to the humanist view, as presented by Ley in this book (Chapter 8), for example, is the subjectivity of the objective. Objects presuppose subjects that have imposed meanings upon them in accord with their individual values, intentions and interests. Intervening between the objective world and behavior is consciousness. However, unlike the marxist view, the humanist is unwilling to acknowledge that just as objects presuppose subjects so subjects presuppose objects. In other words, values presuppose some object to be valued. The desire to eat fish is unlikely to be present among nomads of the Sahara Desert. Or, to be a little more to the point, just as wage laborers cannot be wage laborers without capital so capital cannot be capital without wage laborers.

Indeed, it may be, as is suggested in Chapter 10 of this book, that the humanist view represents much less of an epistemological challenge to behavioral geography than marxism. The reason for this is that the ultimately necessary recognition of the objective presuppositions by humanists can lead in only two directions. One is towards marxism and the other is back into mainstream geography with its positivistic epistemology. Indeed, Ley is already looking forward to 'a synthesis . . . between divergent positions, a synthesis that appropriately links . . . human intentionality and structural factors,' which sounds awfully like the resurrection of subject–object separation.[43]

That behavioral geography is at a critical juncture should now be apparent and, indeed, goes far towards explaining why we believe a re-evaluation to be necessary. It seems likely, on the basis of the evidence of the last five years, that the epistemological challenge from the left will remain just that. On the other hand, even those whom the challengers see as political reactionaries cannot ignore the growing social interdependence of the times; this is an interdependence, moreover, that as a result of government intervention is steadily becoming increasingly apparent. Consequently it seems more likely that behavioral geography will evolve in the directions foreseen by Wolpert and Hägerstrand rather than succumb to an epistemological barrage from without.

Notes

1 Among these were reviews by: Thomas Wilbanks (1971) in the *Annals of the Association of American Geographers*, 61, 618–19; Christine Leigh (1970) in *Environment and Planning*, 2, 482–3; and Allan Frey (1971) in the *Geographical Journal*, 137, 104–5.

2 It should be made clear that, in our view, the program of research endeavor envisaged by behavioral geography never involved the creation of a new branch of the discipline. Rather it was concerned with the elaboration of a distinctive approach to developing theory and solving problems in a wide variety of substantive areas in human geography. Whether the behavioral-geography project succeeded or failed, therefore, depends upon whether or not it was effective in advancing knowledge in those substantive domains, rather than upon

the development of a new coherent subfield within geography.

3 Note, however, that Harvey still describes himself in the Johns Hopkins catalog as a 'behavioral geographer.'

4 T. Hägerstrand (1965) 'On the Monte Carlo simulation of diffusion,' *European Journal of Sociology*, 6, 43–67.

5 J. Wolpert (1965) 'Behavioral aspects of the decision to migrate,' *Papers of the Regional Science Association*, 15, 159–69.

6 D. W. Harvey (1970) 'Behavioral postulates and the construction of theory in human geography,' *Geographica Polonica*, 18, 27–45.

7 G. Olsson and S. Gale (1968) 'Spatial theory and human behavior,' *Papers of the Regional Science Association*, 21, 229–42.

8 A. Pred (1967) *Behavior and Location, I and II*, Lund, Gleerup.

9 The spatial diffusion literature contains numerous examples of such models. See, for example, L. A. Brown (1968) *Diffusion Dynamics: A Review and Revision of the Quantitative Theory of the Spatial Diffusion of Innovation*, Lund, Gleerup, and Hägerstrand, 'On the Monte Carlo simulation' (see note 3 above).

10 The mental map literature is an outstanding example of such studies. See, for instance, P. R. Gould (1973) 'On mental maps,' in R. M. Downs and D. Stea (eds) *Image and Environment*, Chicago, Aldine, Chapter 11.

11 Consider, for example, R. W. Kates (1962) *Hazard and Choice Perception in Flood Plain Management*, University of Chicago, Department of Geography, Research Paper 78; and T. F. Saarinen (1966) *Perception of Drought Hazard on the Great Plains*, University of Chicago, Department of Geography, Research Paper 106.

12 Hägerstrand, 'On the Monte Carlo simulation' (see note 3 above).

13 G. Rushton (1969) 'Analysis of spatial behavior by revealed space preference,' *Annals of the Association of American Geographers*, 59, 391–400.

14 P. Haggett, A. Cliff and A. Frey (1977) *Locational Models*, New York, Halsted Press, 35–6.

15 N. Thrift (in press 1981) 'Behavioral geography: paradigm in search of a paradigm,' in R. J. Bennett and N. Wrigley (eds) *Quantitative Geography in Britain: Retrospect and Prospect*, London, Routledge & Kegan Paul.

16 F. S. Chapin and S. F. Weiss (1962) *Factors Influencing Land Development*, University of North Carolina, Institute for Research in Social Sciences; F. S. Chapin and H. C. Hightower (1966) *Household Activity Systems: A Pilot Investigation*, University of North Carolina, Institute for Research in Social Sciences; F. S. Chapin and R. K. Brail (1969) 'Human activity systems in the United States,' *Environment and Behavior*, 1, 107–30.

17 T. Hägerstrand (1970) 'What about people in regional science?' *Papers of the Regional Science Association*, 24, 7–21.

18 Another major area of difference between the two countries lies in that of cognitive mapping. In the UK this research has until recently progressed little beyond the 'preference surface mapping' pioneered by

Gould under the title 'mental maps' (see note 10 above). The extensive theoretical, philosophical and methodological examinations of these topics that have been pursued in the USA (see R. G. Golledge (in press 1981) 'A critical comment on Guelke's uncritical rhetoric,' *Professional Geographer*, 33, for a review of these) have not been matched by researchers in the UK, at least in part because many of those originally interested in behavioral research were also 'quantitative geographers who have since moved on to other interests or who have retained only a partial interest in behavioral geography. . . .' (Thrift, 'Behavioral geography' (see note 15 above)).

19 L. A. Brown and E. G. Moore (1970) 'The intraurban migration process: a perspective,' *Geografiska Annaler*, B, 52.

20 Wolpert, 'Behavioral aspects of the decision to migrate' (see note 3 above).

21 P. R. Gould (1965) 'Wheat on Kilimanjaro: the perception of choice within game and learning-theory frameworks,' *General Systems Yearbook*, 10, 157–66.

22 Hägerstrand's 'Monte Carlo simulation' conceptualizes the adoption of innovations as a result of a learning process (see note 3 above).

23 R. G. Golledge, R. Briggs and D. Demko (1969) *Configurations of Distance in Intra-Urban Space*, Ohio State University, Department of Geography, Unpublished manuscript.

24 Brown, *Diffusion Dynamics* (see note 9 above).

25 D. R. Reynolds and J. C. Archer (1968) *An Inquiry into the Spatial Basis of Electoral Geography*, University of Iowa, Department of Geography, Discussion Paper 11; K. R. Cox (1969) 'The voting decision in a spatial context,' *Progress in Geography*, 1.

26 G. Tornqvist (1968) 'Flows of information and the location of economic activities,' *Geografiska Annaler*, B, 10, 99–107.

27 E. C. Tolman (1948) 'Cognitive maps in rats and men,' *Psychological Review*, 55, 189–208.

28 R. M. Downs and D. Stea (eds) *Image and Environment*, see note 10 above.

29 G. Olsson (1971) 'Some notes on geography and social engineering,' University of Michigan, Department of Geography, Unpublished manuscript.

30 J. Wolpert (1964) 'The decision process in a spatial context,' *Annals of the Association of American Geographers*, 54, 537–58.

31 J. Wolpert and R. Ginsberg (1969) 'The transition to interdependence in locational decisions,' in K. R. Cox and R. G. Golledge (eds) *Behavioral Problems in Geography: A Symposium*, Evanston, Ill., Northwestern University, Department of Geography, Studies in Geography 17, 72.

32 Hägerstrand, 'What about people in regional science?' 11 (see note 15 above).

33 ibid., 11.

34 ibid., 16.

35 ibid., 14.

36 A. Pred (1973) 'Urbanization, domestic planning problems and Swedish geographic research,' *Progress in Geography*, 5; A. Pred (1977) 'The choreography of existence: comments on Hägerstrand's time geography and its usefulness,' *Economic Geography*, 53, 207–21.

37 For example, A. Buttimer (1972) 'Social space and the planning of residential areas,' *Environment and Behavior*, 4, 285; and D. Mercer (1972) 'Behavioral geography and the sociology of social action,' *Area*, 4, 48.

38 Pred, 'The choreography of existence,' 210 (see note 36 above).

39 J. Wolpert (1970) 'Departures from the usual environment in locational analysis,' *Annals of the Association of American Geographers*, 60, 220–9.

40 D. W. Harvey (1973) *Social Justice and the City*, London, Arnold, Chapter 2.

41 ibid., Chapters 4–7.

42 J. N. Entrikin (1976) 'Contemporary humanism in geography,' *Annals of the Association of American Geographers*, 66, 625.

43 D. Ley (1978) 'Social geography and social action,' in D. Ley and M. S. Samuels (eds) *Humanistic Geography: Prospects and Problems*, Chicago, Maaroufa, 52.

PART 1

1 Inference problems in locational analysis

Gunnar Olsson

Introduction

Contributions to this and other recent volumes indicate that some quantitative geographers have shifted their attention from the modeling of large-scale aggregates to studies of group and individual behavior. As a result, the earlier stress on the geometric outcome of the spatial game has lessened in favor of analyses of the rules which govern the moves of the actors who populate the gaming table. Thus, the new studies aim at a better understanding of those cause-and-effect relationships which are relevant to the decision makers themselves, i.e. to those whose actions eventually will determine the success of various planning programs. With such pragmatic planning ideas in mind, the behaviorists wish to complement the traditional work in quantitative geography by establishing explicit linkages between individual behavior and spatial patterns. Restated and simplified, the behavioral approach suggests a different solution to the geographical inference problem of form and process; while the spatial analyst attempts to infer individual behavior from knowledge of a given spatial pattern, the behaviorist argues for reasoning the other way around.

To assess the epistemological merits of the suggested solutions to the geographic inference problem, the initial purpose of this paper is to discuss current location theories from the viewpoint of the philosopher of science. The secondary purpose is to investigate how well the ideal approach of the behaviorists outlined in the first half of the paper actually compares with their operationalized models. The latter investigation is prompted by the realization that it is one thing to make attractive methodological statements but quite another to translate such predilections into testable formulations.

The truth status of geographic theory

A major proposition of the Vienna Circle is that scientific statements become lawlike by being logically consistent and empirically true, i.e. by being acceptable in terms of both syntax and semantics. A theory may then be defined as a set of deductively connected laws.[1] It follows that once the syntax of a specific theory is accepted as true, its additional value can be assessed in terms of semantics or by its ability to predict empirical events. To test a statement derived from a theory is therefore to seek the instantiation of what is presently considered a law.[2] An important vehicle in this verification procedure is the notion of correspondence rules by which the calculus of the theory can be interpreted in terms of real-world observations, or, if an alternative view is adopted,[3] the calculus of the real world can be interpreted in terms of the theory. In case the correspondence rules cannot be specified or if their application indicates differences between the theoretical and observational languages, then the truth status of the theory is in question. Conversely, a theory is held to be true when all its extralogical terms have factual reference.[4]

Even though geography has rarely been discussed within the rather stringent framework of epistemology,[5] it may still be suggested that predictions based on geographic theory tend to be dubious. This is certainly the opinion of Pred, who argues that the breakdown of location theory is due to discrepancies between the motives which govern the decisions of the actors in the theory and the actors in reality.[6] Others have conveyed essentially the same message about the validity of existing theories but in a lower key and relating it to other causes. For instance, Curry has criticized Lösch's notion of the unbounded plain so severely that he advocates that 'little remains of existing theory to allow its refashioning.'[7] Dacey, finally, is less definitive, despite his claim that it is 'inconceivable that any pattern of central places corresponds exactly to the specified geometry.'[8]

There can be many reasons why theoretical predictions do not sufficiently agree with empirical observations. One reason relates to the aspiration level or the degree of generality at which a particular theory aims. This observation draws on the fact that theories derived within a hypothetico-deductive system by definition can be ordered into a hierarchical structure of statements.[9] An important characteristic of those structures is that axiomatic statements on one level may be testable theorems of a higher-level theory. This suggests that a particular theory should not be rejected simply because it contains unrealistic axioms and therefore may provide unsatisfactory high-level predictions. At the same time, it should be noted that the possibility of syntactical mistakes cannot be ruled out until the theory has been completely axiomatized. It reflects some of the biases of the work in theoretical geography that the only attempts at axiomatization relate to the geometric properties of Christaller's and Lösch's theories.[10]

The previous arguments suggest that the low predictive power of geographic theories perhaps may be due to weak linkages in the interlocking system of hierarchically ordered statements. If this is true, then it is difficult, perhaps impossible, to say anything conclusive about the validity of existing constructs. Since the theories have not yet been fully formalized, it is uncertain exactly which aspiration level is being sought. The root of the uncertainties is supposedly in the employed axioms or perhaps rather in ambiguities introduced when these were provided with semantical meaning and turned into assumptions. It is the purpose of the next section of the paper to discuss the validity of this supposition.

The assumptions of geographic theory

It is sometimes helpful to make a superficial distinction between the spatial and the behavioral axioms of location theory. More exactly, the former postulates relate to the properties of the area over which the actions occur, while the latter concern the motives and behavior attributed to the actors themselves. The behavioral assumptions are usually the same as those of the theory of the firm, while the spatial assumptions commonly are those of the unbounded homogeneous surface. By combining spatial and behavioral postulates, theorems like the hexagonal arrangement of central places may be derived.

To point out that the assumptions of location theory are unrealistic is almost trite. It is far more important that the lack of realism becomes critical only above a certain aspiration level, i.e. when a higher-level hypothesis is refuted by observations that would not refute a lower-level hypothesis.[11] Taking the pragmatic view of many economists,[12] it is enough, therefore, to decide whether the assumptions lead to sufficiently accurate predictions for the purpose at hand – it would be foolish to stop making everyday predictions on the basis of Newton's laws just because of Einstein's subsequent work. Likewise, provided the location analyst is content with devoting himself to pattern analysis and interpretive descriptions of large-scale spatial regularities, he may possibly be satisfied with existing constructs. If, on the other hand, the aspiration level were changed to include analyses also of those micro-units, groups or individuals whose actions give rise to the large-scale regularities, then the situation would be different. The reason is, of course, that the axioms in the first case have become testable theorems in the second. There is little doubt that on this new aspiration level, the axioms of the traditional theory are unacceptable.[13] Instead, the essential problem seems to become that of understanding the internal structure of goal conflicts and adaptive systems. It is for reductions to this level of explanation, i.e. to the level at which inner and outer environments are treated as adapting to one another,[14] that some behaviorally oriented geographers are striving.

The question now arises as to whether the imagined new breed of geographer really is new or whether it is only the old one dressed up as Hans Christian Andersen's Emperor. It would probably seem so to geometrically inclined students like Hudson, who recently claimed that 'one central problem of geographic theory is that of relating individual behavior to that of [spatial] distribution.'[15] Few would object. The difference between the spatial and behavioral approaches to this geographic inference problem becomes clear, however, when Hudson later refers to a 'given system of nodes [within which] the individual must acquire a set of spatial relations so as to navigate . . . in an efficient manner.' Thus, the prime concern is not to derive a spatial pattern from axiomatized behavior but rather to make inferences about behavior from the knowledge of spatial patterns.

Since there are no clearcut one-directional cause-and-effect relationships between geographic form and process, the spatial analysis approach of Hudson and others will almost certainly provide valuable insights. Nevertheless, it is significant that most geographic geometricians have found it necessary to tamper with the classical spatial assumptions and work in transformed non-Euclidean space.[16] Most importantly, the choice between different assumptions is conceived more as a matter of expediency than as a problem involving explicit statements of aspiration level or purpose. This view has been stated most succinctly by Tobler, who asserts that the theory can be made 'more realistic by relaxing the assumptions, but [that] this generally entails an increase in complexity.' Difficulties arise later, however, when this proposition is being executed through the removal of 'the differences in geographic distribution by a modification of the geometry or of the geographical background.'[17]

The terms 'geographic distribution' and 'geographical background' may convey the impression of areal variations in physical landscape and population densities, i.e. of phenomena that have nothing or very little to do with individual or group behavior. Judging from the rest of the paper, however, Tobler must have something more far reaching in mind. Thus, the idea is illustrated by references both to the logarithmic migration maps of Hägerstrand[18] and to cartograms like 'A New Yorker's Idea of the United States of America', i.e. to maps explicitly derived from the theory of cognitive behavior. This means, of course, that the aspirations in fact have been increased to a level where the spatial axioms take on the role of testable theorems derivable from the theory of cognitive behavior.

What is crucial in the comments on map transformations is not that people seem to do something they claim not to be interested in. Instead, it is the consequences that the approach has for the testing of classical location theory. More exactly, it remains to be seen how the data underlying the estimation of transformation functions can be separated from the empirical information with which the theoretical predictions are to be compared; clear specifications of the employed correspondence rules should help to clarify this issue. Even then, however, it may be questioned whether the approach actually is as expedient as sometimes suggested. Speaking exactly to this point, Curry has recently observed that 'if we could gain the level of sophistication necessary to transform, we could probably write theory in terms which would not require it.'[19]

In short, it appears unclear which exact linkages in the alleged chain of deductive reasoning suggested that map transformation would be a valid approach. In less clever accounts than the ones by Tobler, it even seems that large-scale data occasionally have been used as a basis for inferences about small-scale behavior. It is tempting to speculate that such logical peculiarities stem either from the traditional reliance on the map as a given, or from the almost metaphysical belief that the same model can be applied to both physical and human phenomena.[20] More importantly, though, the previous discussion suggests the spatial postulates of location theory to be special cases of behavioral theorems; the questions asked by the spatial and behavioral analysts therefore tend to belong to different scientific aspiration levels. This suggestion supports the commonsense conclusion that the reliability, explanatory power and the potential planning applications of any social-science theory depend on its treatment of individual and group behavior.

Since the behavioral assumptions of location theory recently have been treated elsewhere, it now seems superfluous to discuss those in the same detail as the spatial assumptions.[21] Likewise, it is rather pointless to elaborate on the lack of realism in terms of actual decision processes. Notwithstanding, it is possible that the postulates will still lead to sufficiently good results, provided that only large-scale predictions are aspired to. If, on the other hand, the goal is to understand the finer workings behind large-scale regularities, then it is doubtful whether the traditional approach with its firm grounding in classical utility theory and normative economics will provide reasonable explanations. Examples of other, supposedly more fruitful approaches, include Hägerstrand's ideas about information diffusion and migrations, Curry's notions of shopping lists, inventories, queuing and central places, Wolpert's work on the spatial attributes of stress and goal conflicts, and Pred's explorations into the behavioral matrix.[22] The obvious conclusion is that large-scale patterns should be deduced from explicit statements about individual behavior rather than the other way around. More specifically, spatial patterns should be viewed as reflections of habits and institutionalizations, which in turn can be accounted for by individual decisions governed by continuous learning processes.

The attempts to derive spatial patterns from realistic assumptions about individual or group behavior are appealing because they aim at higher-level explanations. Occasionally, however, some issues relating to the axioms, theorems and aspiration levels still remain unclear. For instance, Curry has recently suggested that the problem in writing theory is to obtain postulates that are not so directly linked to the final results that added insight is not gained.[23] Provided this means that one should not use the same data sets for estimating parameter values as for testing theoretical predictions, or that logical deductions may lead to the discovery of new types of scientific facts, then there can be no disagreement. If, on the other hand, the term 'gaining added insight' refers to something less tangible, then the statement becomes less clear. This is particularly so in view of the tautological nature of the deductive method.[24]

Inference from geographic models

The discussion thus far has centered on some epistemological issues relevant to the geographic inference problem of how to connect spatial patterns and human behavior. It has been proposed that this basic problem of form and process can be approached from two basically different directions. Thus, one may arrive at conclusions about individual behavior through analyses of given spatial patterns, or one may draw conclusions about spatial patterns from detailed knowledge of individual behavior. Both approaches involve difficult inference problems, some of which are related to the choice of axiomatic system or aspiration level, while others are embodied in the lack of one-directional cause-and-effect relationships. Given this situation, continued epistemological re-evaluations of locational analyses seem mandatory.

Evidence that such re-evaluations are important is provided by the fact that even the very superficial comments of this paper have helped to isolate some attractive features of the behavioral approach. The question therefore arises as to whether the logically appealing syntax can be matched by meaningful semantics. To illuminate this question, the discussion will now turn to a review of some models recently employed by students frequently associated with the behavioral school. It should be noted that only operationalized and tested formulations will be treated; to extend the comments into suggested, hypothetical and nontested constructs would not add anything substantive to the comments already made.

As an introduction to the model review, it may be helpful to rephrase the geographic inference problem in terms of large-scale patterns with small variances and small-scale processes with large variances. It follows that large-scale systems, i.e. systems from which portions of the internal variation have been filtered out, tend to be more deterministic, while small-scale systems are more probabilistic. For this reason, the efforts of rewriting well-tested deterministic models like the gravity, regression and rank-size formulations in probability languages becomes interesting from the inference point of view. These attempts to get around rather than solve the problem have been discussed in detail elsewhere,[25] and it is therefore now sufficient to add a reference to some subsequent works on the entropy concept by Wilson.[26]

Provided the same correspondence rules apply in both cases, the rewriting of classical models in probability terms will make large-scale regularities interpretable and derivable from explicit and quasirealistic assumptions about individual or group behavior. However, the value of these reformulations should not be over-estimated; the fact that the final results have been arrived at via another route has not appreciably changed the character of the models. As a consequence, the regulating forces remain deviation counteracting rather than deviation amplifying.[27] On the practical level, however, the translation of the same model to another language with important syntactical and semantical differences may have considerable utility. Thus, it is a common problem in planning situations to determine the scale below which probability techniques must be used; the current discussion of the efficacy of population planning as compared to family planning offers an excellent case in point.[28]

A related but methodologically very different answer to the problem of connecting large-scale spatial regularities with small-scale generating mechanisms is provided by the Monte Carlo simulation technique. Thus, the main characteristic of the simulation technique is that evolving spatial patterns are viewed as resulting from an interplay between deterministic and random factors. More specifically, the general development is determined by distance functions translated via the relative frequency interpretation of probability into the operational form of mean information fields, while the exact development is influenced by a large number of chance factors, operationally represented by the drawing of random numbers. This means, of course, that the resulting patterns may be viewed as a set of 'regulated accidents,' in which chance and contingency have played their game with laws of nature and human behavior.

It has sometimes been suggested that conceiving reality as the result of regulated accidents is paradoxical because the observable events are said to obey the laws of chance, while the underlying probabilities in themselves obey some causal law.[29] On the other hand, the same observation has been extended into the epistemological tenet of complementarity. The issue can be discussed either in terms of axioms and theorems as in the first part of this paper, or in terms of the idea that causal laws at one level of aggregation normally result from averages of statistical behavior at a deeper level, which in turn can be explained by deeper causal behavior, and so on indefinitely.[30]

Diffusion work after Hägerstrand[31] has focused almost entirely on refining details of the original model. With few exceptions, the efforts have been devoted to the construction of specialized computer programs,[32] experiments with different mathematical distance functions,[33] and the derivation of biased or unbiased mean-information fields,[34] while little has been done with the more basic issues of testing and interpretation of underlying theory and functional relationships. This means that most diffusion students – Hägerstrand himself not included – have in fact neglected the behavioral approach to the geographic-inference problem. Thus the employed procedures involve implicit reasoning from the large-scale regularities of the mean-information field to the behavior of the individuals as this is governed by the random-number matrix. It follows that more insight may be gained through detailed experiments with different parameter values based on observed systematic spatial and temporal variations in resistance, distance sensitivity, communication networks and so on. Proceeding to the testing of subsequent model generations, this suggests that more attention should be paid to sensitivity analyses and less to evaluations of spatial end products.

In practice, spatial applications of the Monte Carlo technique rely almost exclusively on large-scale aggregate data, which then are treated as the joint product of deterministic and random variables. As a consequence, inferences about individual behavior can be made only indirectly via a reasoning from spatial patterns to generating mechanisms. The same characterization generally applies to the cell-counting technique,[35] even though Dacey in some of his county-seat models has attempted to deduce spatial distributions from explicit assumptions about underlying processes.[36]

The inferential problem involved in the cell-counting technique is best illustrated in studies employing the negative binomial distribution. It is well known that this distribution can be generated in at least six different ways, some of which are complete opposites. For instance, a spatial point pattern can be described by the negative binomial if it consists of randomly located clusters generated through a two-stage diffusion process, in which the parent points have been distributed randomly over the area and the secondary points have been assigned among the initial nuclei independently of one another but in such a fashion that the growth over time is logarithmic. The opposite to this generating mechanism is the urn scheme for heterogeneous Poisson sampling, according to which a negative binomial may be obtained for the total area, provided the area can be divided into regions within which the points have been randomly distributed but in such a manner that the mean number of points per cell varies between the regions according to the gamma function. The obvious conclusion is that very little can be said about generating mechanisms solely on the basis that the morphology of a point pattern can be described by the negative binomial.[37]

It could perhaps be tempting to conclude from the discussion of the negative binomial that it is possible to reason from process laws to morphological laws but not in the other direction. However, not only are the causal linkages in geography too intricate to allow such a conclusion, but it may also be shown mathematically that at least one single mechanism – that of space filling – can give rise to either clustered, random or regular spatial patterns. In such situations of conflicting results, one solution is to fit the same data to another set of distributions connectable with only one of the previous interpretations.[38] Despite its value in the special case, this approach is clearly more expedient than elegant. It will hardly bring the solution of the geographic-inference problem much closer.

To varying degrees, the models discussed have started from a given spatial pattern and then proceeded to indirect inferences about generating processes and individual behavior. This conclusion has been possible to substantiate only because the cited models have been refined to the extent of operationalization and empirical application. Unfortunately, the rest of the model work in behavioral geography still awaits rigorous testing, and therefore does not permit the same degree of conclusiveness. On present evidence, however, it is not unlikely that current theoretical explorations will end up with testable formulations that are based on the traditional approach to the geographic inference problem rather than on a new and logically more attractive one.

The suspicion that techniques for generating spatial patterns from individual behavior still may be far away relates closely to the attempted adaptations of classical learning theory to the geographers' need.[39] Thus, the focus has been more on how individuals learn to act efficiently in an existing spatial system than on how their actions cause existing spatial patterns to change. Basically, the same holds for the notion of subjective preference functions[40] and for most studies of mental maps.[41] Ignoring the extremely thorny measurement problems as being beside the point in the present context, the most interesting property of the latter studies is their amenability to trend analysis. Perhaps it is on this level that the relationships between the allegedly new approaches and the traditional work in spatial analysis become most evident; trend-surface analyses and map transformations appear in fact to be the dual of one another.

Summary and conclusions

This paper was based on the premise that the limited predictive power of geographic theories is due to a preoccupation with spatial patterns and a neglect of small-scale generating processes. The paper has attempted to evaluate this premise by comparing the spatial and the behavioral approaches to the geographic inference problem of form and process. To establish some general guidelines, attention was first given to the overlaps between geography and the philosophy of the social sciences. The subsequent conclusion was that the behavioral axioms of location theory belong to a higher level of the hierarchical structure of the hypothetico-deductive system than do the spatial axioms. As a consequence, the behavioral approach can provide more detailed explanations and is therefore preferable, particularly if the findings are to be extended into planning applications.

It is one thing, however, to isolate attractive methodological approaches and quite another to translate these predilections into operational and testable models. In order to assess what has actually been achieved rather than merely talked about, the latter half of the paper reviewed a number of models recently used by students more or less identified with the behavioral school. More specifically, attention was given to the rewriting of deterministic models in probability terms, the use of Monte Carlo simulations, the cell-counting technique, and the geographical amendments to psychological learning theory. It was found that practically all studies had started from given spatial patterns and then proceeded to indirect inferences about generating processes and underlying human behavior. Although suggestions about alternative and epistemologically more attractive approaches do exist, it has been difficult to find cases where such models actually have been applied to empirical data. Recalling the positivists' quest for combinations of logical consistency and empirical truth, this leaves the assessor bewildered. On the one hand, it is possible to point to a number of low-order spatial formulations with considerable empirical reliability. On the other hand, one may imagine some logically attractive behavioristic formulations that unfortunately still await empirical evaluations.

The final conclusion must be that the behavioral approach to quantitative geography may or may not alter the current rather peculiar state of the art. Speaking *for* improvement is the growing recognition of studies from quantitative psychology and non-normative economics as well as the epistemological bases of most behavioral work. Speaking *against* substantial and quick change is the existence of multidirectional causal relationships as well as the shortage of suitable highly disaggregated data.

Notes

1 M. B. Turner (1967) *Philosophy and the Science of Behavior*, New York, Appleton-Century-Crofts, 226; M. Brodbeck (1968) *Readings in the Philosophy of the Social Sciences*, New York, Macmillan, 583.
2 T. S. Kuhn (1962) *The Structure of Scientific Revolutions*, University of Chicago Press.
3 H. Margenau (1950) *The Nature of Physical Reality*, New York, McGraw-Hill; N. R. Hanson (1958) *Patterns of Discovery*, Cambridge University Press.

4 C. G. Hempel (1965) *Aspects of Scientific Explanation*, New York, Free Press, 217–22.

5 For notable exceptions see, F. Lukermann (1961) 'On explanation, model and description,' *Professional Geographer*, 13, 1–5; R. Golledge and D. Amedo (1968) 'On laws in geography,' *Annals of the Association of American Geographers*, 58, 760–74; G. Olsson (1968) *Distance, Human Interaction and Stochastic Processes: Essays on Geographic Model Building*, Ann Arbor, University of Michigan; D. Harvey (1973) *Explanation in Geography*, London, Arnold (new edn in press 1981).

6 A. Pred (1967) *Behavior and Location, I*, Lund, Gleerup.

7 L. Curry (1962) 'The geography of service centers within towns: the elements of an operational approach,' in K. Norborg (ed.) *Proceedings of the IGU Symposium in Urban Geography*, Lund, Gleerup, 33.

8 M. F. Dacey (1964) 'Imperfections in the uniform plane,' Michigan Inter-University Community of Mathematical Geographers, Discussion Paper 4, 1.

9 R. B. Braithwaite (1960) *Scientific Explanation*, New York, Harper.

10 M. F. Dacey (1965) 'The geometry of central places,' *Geografiska Annaler*, B, 47, 111–24.

11 Braithwaite, *Scientific Explanation* (see note 9 above).

12 M. Friedman (1953) *Essays in Positive Economics*, University of Chicago Press.

13 H. Simon (1959) 'Theories of decision-making in economics and behavioral science,' *American Economic Review*, 49, 253–83; R. M. Cyert and J. G. March (1963) *A Behavioral Theory of the Firm*, Englewood Cliffs, N.J., Prentice-Hall.

14 H. Simon (1969) *The Sciences of the Artificial*, Cambridge, Mass., MIT Press.

15 J. Hudson (1969) 'A model of spatial relations,' *Geographical Analysis*, 1.

16 W. Bunge (1966) *Theoretical Geography*, Lund, Gleerup.

17 W. R. Tobler (1963) 'Geographical area and map projections,' *Geographical Review*, 53, 59–78.

18 T. Hägerstrand (1957) 'Migration and area,' in D. Hannerberg, T. Hägerstrand and B. Odeving (eds) *Migration in Sweden*, Lund, Gleerup.

19 L. Curry (1967) 'Quantitative geography 1967,' *Canadian Geographer*, 11, 265–79.

20 M. J. Woldenberg (1968) 'Spatial order in fluvial systems: Horton's Laws derived from mixed hexagonal hierarchies of drainage basin areas,' *Bulletin of the Geological Society of America*, 80, 97–112.

21 G. Olsson and S. Gale (1968) 'Spatial theory and human behavior,' *Papers of the Regional Science Association*, 21; D. Harvey (1970) 'Behavioral postulates and the construction of theory in human geography,' *Geographica Polonica*, 18.

22 T. Hägerstrand (1953) *Innovations Förloppet ur Korologisk Synpunkt*, Lund, Gleerup; L. Curry (1967) 'Central places in the random

economy,' *Journal of Regional Science*, 7, 217–38; L. Curry (1969) 'A "classical" approach to central place dynamics,' *Geographical Analysis*, 1; J. Wolpert (1966) 'Migration as an adjustment to environmental stress,' *Journal of Social Issues*, 22, 92–102; Pred, *Behavior and Location* (see note 6 above).

23   Curry, 'Central places in the random economy' (see note 22 above).

24   Hempel, *Aspects of Scientific Explanation* (see note 4 above).

25   G. Olsson (1967) 'Central place systems, spatial interaction and stochastic processes,' *Papers of the Regional Science Association*, 18, 13–45.

26   A. G. Wilson (1968) 'Notes on some concepts in social physics,' Centre for Environmental Studies, London, Working Paper 4.

27   W. Buckley (1967) *Sociology and Modern Systems Theory*, Englewood Cliffs, N.J., Prentice-Hall.

28   K. Davis (1967) 'Population policy: will current programs succeed?' *Science*, 158, 730–9.

29   M. Born (1964) *Natural Philosophy of Cause and Chance*, New York, Dover.

30   W. Kneale (1968) 'Scientific revolution for ever?' *British Journal for the Philosophy of Science*, 19, 27–42.

31   Hägerstrand, *Innovations Förloppet* (see note 22 above).

32   F. R. Pitts (1967) 'MIFCAL and NONCEL: two computer programs for the generalization of the Hägerstrand models to an irregular lattice,' University of Hawaii, Social Science Research Institute, Working Paper 4.

33   R. L. Morrill (1963) 'The distribution of migration distances,' *Papers of the Regional Science Association*, 11, 75–84.

34   D. F. Marble and J. D. Nystuen (1963) 'An approach to the direct measurement of community mean information fields,' *Papers of the Regional Science Association*, 11, 99–109; R. L. Morrill and F. R. Pitts (1967) 'Marriage, migration and the mean information field: a study in uniqueness and generality,' *Annals of the Association of American Geographers*, 57, 401–22.

35   D. Harvey (1968) 'Geographic processes and the analysis of point patterns: testing models of diffusion by quadrat sampling,' *Transactions of the Institute of British Geographers*, 44, 85–95.

36   M. F. Dacey (1966) 'A county seat model for the areal pattern of an urban system,' *Geographical Review*, 56, 527–42.

37   W. Feller (1943) 'On a general class of contagious distributions,' *Annals of Mathematical Statistics*, 14, 389–400; D. Harvey (1968) 'Some methodological problems in the use of the Neyman type A and the negative binomial probability distributions for the analysis of spatial point patterns,' *Transactions of the Institute of British Geographers*, 44, 85–95.

38   G. Olsson (1968) 'Complementary models: a study of colonization maps,' *Geografiska Annaler*, B, 50, 1–18; R. G. Swinburne (1969) 'Vagueness, inexactness and impression,' *British Journal for the Philosophy of Science*, 19, 281–99.

39 R. G. Golledge (1967) 'Conceptualizing the market decision process,' *Journal of Regional Science,* 17, 239–358; R. G. Golledge (1969) 'The geographical relevance of some learning theories,' in K. R. Cox and R. G. Golledge (eds) *Behavioral Problems in Geography: A Symposium,* Evanston, Ill., Northwestern University, Department of Geography, Studies in Geography 17; R. G. Golledge and L. A. Brown (1967) 'Search, learning and the market decision process,' *Geografiska Annaler,* B, 49, 116–24.

40 G. Rushton (1969) 'The scaling of locational preferences,' in Cox and Golledge, *Behavioral Problems* (see note 39 above).

41 P. R. Gould (1973) 'On mental maps,' in R. M. Downs and D. Stea (eds) *Image and Environment,* Chicago, Aldine; P. R. Gould (1968) 'Problems of space, preference measures and relationships,' *Geografiska Annaler,* B, 50; P. R. Gould and R. R. White (1968) 'The mental maps of British school leavers,' *Regional Studies,* 2, 161–82; R. M. Downs (1967) 'Approaches to and problems in the measurement of geographic space perception,' University of Bristol, Department of Geography, Seminar Papers, A, 9.

2 Conceptual and measurement problems in the cognitive–behavioral approach to location theory

David Harvey

In this paper I want to examine some of the problems that arise from taking a cognitive–behavioral approach to location theory. The argument for this approach may be summarized as follows. Locational patterns in human geography are the physical expression of individual human decisions. Locational analysis must therefore incorporate some notions regarding human decision making. The simplest course is to set up idealizations or to develop some descriptive device to summarize aggregate human behavior. The idealization of rational economic man leads us to the normative location models such as those of Weber, von Thünen, Lösch and their academic descendants. Empirical evidence suggests that it is possible to conceptualize behavior as a stochastic decision process and use probability distributions to discuss spatial behavior, provided we are considering basically similar choices in a fairly homogeneous population. Descriptive mathematical functions may then be used as the foundation for a stochastic location theory – a theory that has yet to be written and the characteristics of which remain largely unknown in spite of recent general formulations.[1] But there are many situations in geography in which these descriptive devices or idealizations are clearly inappropriate and there is no alternative but to incorporate very specific statements about the cognitive processes involved in the act of decision.

We know that decisions are affected by attitudes, dispositions, preferences and the like. We know, too, that mental processes may mediate the flow of information from the environment in such a way that one individual perceives a situation differently from another even though the external stimuli are exactly the same. Each individual may be thought of as making decisions with respect to his attitudes and in the context of his perceptions. We also know that an individual's attitudes and dispositions may be affected, often cumulatively over time, by the constant bombardment of stimuli from the environment around him and by cultural conditioning.

We can perhaps abstract the sense of a cognitive–behavioral location theory as follows. It should be able to handle a process in which each individual decision maker, enclosed in his own environment, reaches a decision which presumably maximizes some 'satisfaction' or 'preference' function defined over his own dispositions and attitudes. To be a viable theory it must also be able to handle problems of aggregation that result either from any macrolocational analysis that is required, or from needing to resolve the problem of conflicting decisions in complex organizational structures. It would, for example, be unthinkable to discuss organizational decisions, such as those of government, without reference to the problems of conflict resolution. In addition the theory will need to handle the feedback effects from the environment to the decision maker. The simplest example of this kind of feedback is the learning process which results when stimuli from the environment reinforce the attitudes and dispositions of the decision maker.

At first sight the formulation of such a cognitive–behavioral location theory appears an attractive but formidable task. It is attractive because it seeks to understand the decision process as it really is. I suspect that it also has intuitive appeal for many because it satisfies hopeful and hidden emotions about freedom of choice, individuality, and, ultimately, free will. In this respect it seems to function as a new and more sophisticated version of a dodo that refuses to die in geographic thought – the notion that everything really *is* unique. Formulation of such a theory appears formidable because, if it is ever to be anything more than a vague hopeful speculation, it will need to settle a whole host of conceptual and measurement problems in such a way that we can actually understand what has eluded the behavioral sciences as a whole – viz., the *real* reasons why people behave with respect to their environment in the way that they do. It is unlikely, of course, that we will achieve complete understanding. The question therefore, is not whether we will construct a theory to explain everything about human decision making, but how far and how quickly we can progress along this road, constructing reasonable partial formulations as we go. It is useful to consider this question if only from the point of view of research strategy. If, as is possible, it will take enormous research effort and time to construct quite trivial behavioral formulations, then it might be advisable to shift our intellectual resources elsewhere. We thus find ourselves in the situation of the decision maker in the face of uncertainty. We cannot know the final answers until we have tried all possible strategies. But like all decision makers in the face of uncertainty we can generate certain expectations and make our decisions with respect to them.

It seems to me that we have three strategies open to us.[2] We can seek to extend 'classical' location theory with its emphasis upon optimization techniques; we can seek to build a stochastic location theory; or we can take a cognitive–behavioral approach. These are not mutually exclusive or easily separated strategies. They should be thought of as three different focal points in the universal set defined by the decision problem. It is useful to ask, however, what we might expect the ultimate relationships to be between the theories generated around these three foci. My own expectation is that these theories will have distinctive and only partially overlapping domains. They will be complementary rather than competitive. When policy issues regarding economic efficiency are involved, normative economic location theory cannot be replaced although there is, of course, plenty of room for its improvement. There are undoubtedly many situations in which, either through the forces of competition or through an inherent tendency among decision makers to seek out optimal or close-to-optimal solutions, the normative economic theory or extensions of it will provide us with a reasonable and quite handy model for rather more complex decision processes. This is not, perhaps, a fashionable view among geographers at present. It does, however, seem to me to be a totally unwarranted inference that normative economic location theory has no empirical relevance because we can so frequently find deviations between patterns predicted from theory and actual patterns. The trouble with normative location theory is not that it is empirically irrelevant, but that we do not know the circumstances in which it may be used as an empirical device. It is uncontrollable rather than irrelevant in empirical work. Indeed, one of the side payoffs from formulating an adequate cognitive–behavioral location theory may be to improve our control over the empirical use of normative models. I think it also reasonable to expect that stochastic models of behavior will not be challenged by a cognitive–behavioral theory in certain domains. Particularly when we are concerned with the aggregate effects of countless individual decisions (about which it is very difficult to collect any definite information), it would seem senseless to attempt a behavioral analysis of each individual decision and then try to aggregate these up into a model that copes with the total process. Probability theory, especially when given a relative frequency interpretation, provides a set of extraordinarily effective models for dealing with aggregate effects of decisions that are rather repetitive in form over time and space. The general aggregative characteristics of migration, journey-to-work, journey-to-shop, diffusion, and so on, can and will most easily be handled by the use of stochastic models. Considering the well-documented statistical regularities which can be observed in human behavior in space, it appears very reasonable to expect the integration of many of these concepts into some basic stochastic location theory. We are then left with a 'residual' domain of events which cannot be handled by the normative or stochastic location theories.

This domain will presumably be colonized by a cognitive–behavioral location theory. Let us consider the kind of event which this domain might contain. There may, for example, be policy problems in which criteria of economic efficiency are either difficult to define or irrelevant, or situations in which the goals of societies, groups, or individuals are clearly non-economic. A cognitive–behavioral theory will need to handle the complex problems of value judgements, utility scales, and so on. There may also be situations in which a very few decision makers have a disproportionate effect upon spatial patterns. At the microlevel we may be concerned with how an individual or a small sample of individuals is acting in a given situation (for example, a dozen or so farmers in a particular area). In this case as the sample size increases so the stochastic models may become more relevant. At the macrolevel we may be concerned with the decisions of a few individuals in government or big business since these decisions can have an enormous effect upon regional development. In these circumstances it is vital to understand the attitudes and dispositions (particularly political dispositions) of those involved in the decision process, and to understand how alternatives are perceived, searched, selected and implemented. The domain of a cognitive–behavioral location theory will thus range from crosscultural variation in value judgements and perceptions through individual choice behavior to group decision-making processes.

It is pertinent to speculate now about the prospects for formulating reasonable theories in the cognitive–behavioral domain. Again, we are forced to speculate in the face of uncertainty, but certain expectations can be generated to help make decisions on research strategy. I rest these expectations on my knowledge (which is obviously incomplete) of the conceptual and measurement apparatus currently available to us. Future research results will undoubtedly alter the picture considerably, but since these are unknown it seems best to base our assessment on current information. It is useful to concentrate on conceptual and measurement apparatus because provision of these is a *sine qua non* for successful theoretical formulations and their application.

Concepts provide us with analytic power, while measurement procedures provide us with the necessary techniques to pin down our analyses to the world of experience. Concepts and measurement techniques are not independent of each other. Strict operationalists would claim, for example, that every measurement procedure must logically bear a unique relationship to concept. They would also argue that each concept can only be defined by reference to the procedures that are used to gain knowledge of it – and these procedures often involve measurement. I do not wish to take such a strict operationalist view in this chapter. The link between concepts and measurement procedures is undeniable, but we can usefully separate them and on occasion regard them as being quite different from each other.

Concepts in general

I take it as axiomatic that we can hope to handle the complex world of behavior only by formulating firm concepts with generally agreed meanings. If we are to communicate our ideas we must first agree upon some way of assigning meaning to terms and this amounts to agreeing upon some procedure of definition. To do this we must define concepts and relate them to experience.

It is worth distinguishing between *theoretical* concepts and *empirical* concepts. This distinction is important because the assignment of meaning differs radically according to type, and because it is important not to mix the two. Theoretical concepts can be defined implicitly. Empirical concepts can be defined explicitly. To understand this better we need to define our own terms and we can best begin by asking what we mean by the term theory. I shall simply regard it as an abstract calculus in which the terms can be defined by their syntactical function rather than by reference to their empirical interpretation. This is perhaps a difficult idea to grasp, but it can be illustrated most easily by mathematical examples. The terms point and set in mathematical formulations have no empirical meaning, and if we wish to give them meaning we can do so by examining the way they function in the calculus. It is rather like defining the rook in chess by specifying the field of play and the rules governing its movement. Now it is possible to relate these theoretical concepts to perceptual experience by establishing a set of 'correspondence rules' or 'epistemic correlations'[3] between them and empirical concepts. A mathematical point may thus be correlated with a dot on a piece of paper. This epistemic correlation does not, however, exhaust the possible semantic interpretations of the theoretical concept. In the applications of probability theory the same concept of a point, this time located in a sample space, may be interpreted as an outcome of an experiment. Establishing epistemic correlations provides a rather different way of assigning meaning to a theoretical concept. But in this case the assignment is implicit and incomplete. Now, it was the operationalist argument that the only worthwhile manner of assigning meaning was by way of such epistemic correlations. This is now generally regarded as being unnecessarily restrictive and not very advisable because it makes the theoretical structure incapable of further application and further growth.[4]

Empirical concepts are capable of explicit definition with respect to experience. We can provide ostensive definitions or provide operational definitions. It is also a characteristic of empirical concepts that when we give them lexical definitions, we can replace the term being defined by the *definitions* without any loss of information. Thus we can exhaust the meaning of an empirical concept if we so wish by specifying fully the operations by which knowledge of that concept is obtained.

We possess two languages. The language of theory contains

concepts that can be defined syntactically although they can also be interpreted in terms of real-world experience via epistemic correlations. The language of empirical investigation contains concepts which can be defined operationally, explicitly and completely. If theory is to be of any use we need to be able to translate from one language to the other. This is not always easy to do in the social sciences.[5] Empiricists complain that social science theory is in general incapable of empirical interpretation and therefore not worth bothering with. Theoreticians complain that their theories remain uninterpreted because empiricists have failed to establish the kinds of concept that will allow epistemic correlations to be made. There are, of course, many situations in which this translation has successfully been made. In a stochastic location theory, for example, the empirical concept of a town (which can be given an operational definition) can be translated into a mathematical point in a sample space. In most cases these translations run from empirical concepts to mathematics rather than from empirical concepts to social-science theory. This latter form of translation is not helped by the fact that we often use the same term to represent both theoretical and empirical concepts. A point, for example, has meaning in both language systems and it is important to differentiate in which sense it is being used. It might almost be worthwhile to adopt a notational system to make explicit which language we are using when there is any possibility of ambiguity.

Let us now consider some of the concepts available to us in behavioral science in the light of these two kinds of language. There is certainly no shortage of theoretical concepts. Indeed, looking round the behavioral sciences it is difficult to know where to begin. It is easiest to proceed by example. I shall therefore begin by examining two concepts of considerable importance to our argument, *economic rationality* and *satisficing behavior*. The first lies at the very center of economic location theory and the second, judging by its frequency of use in geographic literature, is an important if not central concept in the cognitive–behavioral approach to location theory.

The concept of economic rationality has been an extraordinarily fruitful one. It functions as one of the primitive terms of economic location theory and, like most primitive terms, it can be used to generate derived terms – concepts of marginal behavior, pure and perfect competition, profit maximization, and so on, readily spring to mind. From this it is easy to see that the concept of economic rationality can be given a firm syntactical definition in terms of its function in theoretical economics, which includes, of course, the economic theory of location. The concept of satisficing behavior has a much shorter history and it would be churlish, therefore, to expect it to be as well developed syntactically as the concept of economic rationality. But the fact remains that it has not been a very fruitful concept. In some respects it is theoretically ambiguous. I think the basic trouble with it is that it is a negative rather than a positive concept. It is designed to explain the empirical shortcomings of economic theory rather than to generate theory in its own right. But it is still useful to ask how it might function in a cognitive–behavioral location theory.

The concept of satisficing behavior is a very confused one. It has several connotations. Let us consider, for example, the relationship between satisficing behavior and optimizing behavior. We can, if we wish, regard satisficing behavior as a form of optimizing behavior in which the criteria used are non-economic. Instead of writing an objective function that maximizes profits, for example, we might write one that maximizes leisure time subject to the constraint that a certain minimum income is achieved. There is another possibility. Satisficing behavior may be regarded as optimizing behavior (of any kind) with respect to a number of preselected alternatives out of a much larger (sometimes infinite) set of alternatives. In this case the concept of satisfaction may refer to the decision maker's intuitive assessment of the adequacy of his preselection process. I suspect that this is what Simon actually meant by the term 'bounded rationality.'[6] But there is yet another possible interpretation. Even given a bounded choice and non-economic criteria, the decision maker does not seek *any* optimal solution. In this case satisficing behavior means non-optimizing behavior. I suspect this last interpretation is theoretically barren, yet it is an interpretation that geographers are rather partial to. It is theoretically barren because theories are deductive structures and this interpretation of the concept of satisfaction is only capable of being exploited inductively. In short, if we accept this interpretation almost *any* form of decision-making behavior could follow from it. Now it may be the case that decision making really is characterized by non-optimization in the world of experience – I do not wish to deny this possibility – but a theoretical concept of satisficing behavior is worse than useless if it merely refers to this possibility without giving us any further clue as to how we might handle such situations. My conclusion is that we either need to interpret satisficing behavior as some form of optimizing behavior, or we must abandon the concept and seek for theoretical concepts which do not give us some control over the analysis of non-optimizing behavior. We will consider certain possibilities from perception studies later.

How can these two concepts of economic rationality and satisficing behavior be given an interpretation in empirical language? What kinds of epistemic correlations can be established? In neither case, of course, can these epistemic correlations exhaust the theoretical meaning of the concept, and several epistemic correlations are possible in each case. Again, it seems to me that the concept of economic rationality has been given more and more fruitful empirical interpretations than has the concept of satisficing behavior. There are many empirical situations in which it is possible to define profit-maximizing or cost-minimizing behavior in an operational sense, and most of the techniques of operations research are available for discussing these problems empirically. Operational definitions can be found for concepts of profit, loss, cost and so on. There are many situations in which it is possible to measure exactly, and thereby achieve operational control over, the concepts concerned. In other cases operational definitions are less easy to come by. The theoretical concept of profit, for example, cannot always be given an empirical interpretation. Do we mean short-term or long-term profit, and how long is the long term? How do we measure indirect nonmonetary benefits? What does profit mean in the context of the social system as a whole? Thus the same concept can be given an empirical interpretation in some situations and not in others. I suppose the main objections to the concept of economic rationality (and its derivatives) arise from the failure to discriminate between these two different kinds of situation. But to dismiss the concept because it has been grossly misused is to throw the baby out with the bathwater. If we wish to control the use of the concept of economic rationality in an empirical context we can do so by a careful appraisal of the measurement procedures used to quantify it.

How can the concept of satisfaction be given an empirical meaning? How can we measure it? The answer depends, of course, upon what theoretical interpretations we are giving to the concept. If we regard it as non-optimal behavior (of any kind) then there is only one way we can operationalize it in its present form. We can generate expectations between expected and observed behavior as some measure of the degree of satisficing behavior. Now there seems to be something fundamentally unsatisfactory about this procedure. It assumes the adequacy of the economic optimizing model in the measurement of satisfaction! Deviations between expected and observed behavior may, of course, be explained in a number of different ways. There may be errors of specification in the optimizing model, errors in estimating the parameters, and so on. The deviations are of inherent interest, but there is no guarantee (short of perfection in the optimizing model and in its calibration) that they are realistic measures of satisficing behavior. If, on the other hand, we regard satisficing as a form of optimizing behavior, then it may be possible to operationalize the concept in the same way that we can operationalize the economic model. Consider the various ways in which we might specify an objective function in an operational manner. We may maximize leisure time, minimize labor input, minimize effort, maximize security levels and so on. These can be operationalized. We cannot maximize happiness, joy, or, for that matter, satisfaction, in any operational sense. All we can do, therefore, is to translate happiness or satisfaction into empirical concepts such as leisure time, security levels and so on. As always, the translation is incomplete. But unless we are willing to perform such translations we cannot hope to discuss the concept of satisfaction in empirical terms.

It is worth noticing, however, that theoretical concepts that started out by being polar opposites are now operationalized so that they are different in degree not kind. Both economic rationality and satisficing behavior are interpreted as optimization behavior, and the only difference between them is the nature of the objective function – economic rationality presumes profit is to be maximized, satisficing behavior presumes it is some other quantity, such as leisure time. It could be argued, however, that there is a real difference in the nature of the constraint sets. If we state, for example, that a certain minimum income level must be achieved while maximizing leisure time, then the concept of satisfaction enters into the problem by way of the arbitrary choice of this minimum level. Some decision makers might set it at $10,000 and others at $20,000, and so on. Without this information the model would certainly not be realistic. But this particular arbitrary decision does not strike me as being any different in principle from the kind of arbitrary decision necessary in the economic model regarding the resource constraints (how much labor to use, and so on). It is also worth noting that in neither case is the concept of omniscient understanding employed. In empirical work the concept of profit maximization has never assumed total knowledge. The alternatives evaluated are usually a few out of all possible alternatives. In both cases, therefore, we make a choice from a bounded set. Therefore, if we accept the idea of satisficing behavior as optimizing behavior, the difference between the former and economic rationality narrows very appreciably in the language of empirical research. There is also a considerable payoff to be had from studying theoretically and empirically the relationships between solutions generated by different objective functions. This is not simply an interesting question – it is a vital one. Many of the social problems we currently face can be conceptualized in these terms. Pollution problems might be considered in terms of a conflict between optimization over the total social system and optimal behavior on the part of individuals operating within that system. Conservation problems might be regarded as a conflict between profit maximizing and, say, leisure maximizing. I have not space, however, to launch into a discussion of these kinds of question here.

If satisficing behavior is regarded as non-optimizing behavior then the above argument is unacceptable. The concept becomes a negative one which merely serves to remind us, often quite appropriately, that profit maximization is not everything in life. But the concept is theoretically and empirically barren. It must, therefore, be replaced with positive concepts which bear theoretical fruit and, preferably, which can be operationalized in empirical work. There are plenty of such concepts available to us. I shall therefore again proceed by example and take a more detailed look at the concepts and measurement procedures available for the study of perception. The concept of satisfaction is often associated with the idea that the decision maker proceeds on the basis of his images or perceptions and as we saw at the beginning of this essay, the notion of a difference between the 'perceived world' and the 'real world' is in any case important in the cognitive–behavioral approach to location theory.

Theoretical concepts in perception studies

The concept of perception is itself rich in ambiguity. In some cases the concept is used 'to designate a world view, an outlook on life, or some other very general cognitive product.'[7] This interpretation has little to recommend it from an analytic point of view. It does remind us, however, that there are some extraordinarily interesting problems in phenomenological philosophy and that 'images' of great generality may have tremendous significance to human decision making.[8] But in this very general sense the concept of perception is no better than the general concept of satisfaction. If we are to handle problems we need more precise definitions than this. Even in technical terms, however, the concept has been given quite different interpretations and it is difficult to avoid the conclusion that it often refers to several different processes. Much work is concerned with the physiological aspects of perception. The processes are essentially visual, auditory, and so on, and the emphasis of study is upon those variables that impinge directly upon the senses. This kind of approach is only indirectly relevant to the construction of a cognitive–behavioral location theory. We are more interested in the social aspects of perception. Not, of course, that the two are independent of each other. But the focus of interest is different. Within the social perception field, however, there are several competing frameworks for examining perceptual processes. Geographers have, at various times, been attracted to different formulations. Some refer to gestalt perception studies,[9] some to field-theory frameworks,[10] and so on.[11] It is hardly surprising to find, therefore, that no generally agreed definition of perception can be supplied. *Precise* definition is probably neither necessary nor desirable, and Warr and Knapper regard overemphasis upon precise definition as a danger to be avoided:

> it is far from disconcerting that books on perception cannot open with a complete definition of this concept . . . it is clear that in some sense perception involves an interaction or transaction between an individual and his environment; he receives information from the external world which in some way modifies his experience and behavior. But beyond statements of this order of generality there are few formulations which are universally accepted. Writers with different backgrounds and objectives tend

to emphasize different aspects of the process, so that various approaches are reflected in varying definitions.[12]

It is possible, therefore, to choose our own definition of perception (within limits) according to our objective. In our case the objective is to formulate a cognitive–behavioral location theory. It seems reasonable to conclude that the definition we choose for the concept of perception should be one relevant for the discussion of spatial behavior. Above all, we would prefer to be able to predict spatial behavior on the basis of our understanding of perception.

At this point it is worth introducing a distinction between attitudes and perceptions. Warr and Knapper[13] suggest that attitudes are relatively permanent structures that hold in the absence of any particular stimulus, whereas perceptions are more flexible and transitory and only occur in the presence of a stimulus. Obviously, there are strong interactions between attitudes and perceptions defined in the above manner; attitudes are presumably formed by perceptual experience and, in turn, affect the 'perceptual readiness' of the individual. I think the difference between attitudes and perceptions is particularly useful to us. If we accept the definition of an attitude as a 'learned predisposition to respond to any object in a consistently favorable or unfavorable way,' then we would, I think, be predisposed to think that attitudes have a powerful influence over spatial behavior. But recent work in fact suggests that a person's attitude to an object is *not* a major determinant of his behavior with respect to that object.[14] There is some effect, of course, but this is by no means as strong as usually is assumed. If this is the case, then we can afford largely to ignore the problem of attitudes in seeking for a cognitive–behavioral location theory. Is it possible, then, to make the concept of perception a cornerstone for such a theory? There are several reasons why I think the answer to this question should be positive. We know, for example, that many decisions are 'impulse' decisions; decisions about purchases of goods, migration, investment, and so on, are frequently of this sort. A concept of perception that refer to transitory events under direct stimulus seems much more useful for the analysis of these kinds of decisions than does the concept of an attitude. It would be difficult to discover what proportion of decisions are of an impulse type (rather more than we usually cater for, I suspect), but even in those cases where attitudes are important, it is quite possible that they are so simply through their effect upon perception. In other words, perception as we are here defining it might be regarded as the central node in a network that brings together cognitive processes and environmental stimuli, and which projects to the act of decision. I write 'might' advisedly because, if we are to gain anything from this broad conceptualization, we must develop a firmer theoretical framework for analysis and establish the necessary epistemic correlations to facilitate empirical work. Because it seems to me to be important to avoid too great a gap between the language of theory and research, I shall endeavour to clarify some theoretical problems by taking a close look at empirical concepts of perception and in particular at the measurement procedures which may be used to gain knowledge of them.

Empirical concepts and the measurement of perception

Ogden and Richards long ago remarked that 'perception can only be treated scientifically when its character as a sign-situation is analyzed.'[15] I want to tackle the problem of measuring perception via semiotics (the theory of signs) partly because this provides a coherent framework for examining the measurement problem, but also because it provides us with a convenient way of bridging the gap between theory and empirical research. There is no need to define a sign with any precision; it may be a symbol (such as a word), a photograph, a map, an object, an experience, and so on. Morris,[16] who has pioneered the theory of semiotics, prefers to leave the meaning open and considers the nature of signs in terms of a general sign process. From our point of view the importance of Morris's presentation is the way in which this sign process relates to behavior.

Morris regards a sign process as a five-point relation in which a *sign* sets up in an *interpreter* (in our case the decision maker) the disposition to act in a certain kind of way (called the *interpretant*) to an object or event (called the *signification*) in the *context* in which the sign occurs. Morris goes on to suggest that signification is tridimensional and that it can be correlated with perceptual, manipulatory and consummatory aspects of action:

> The organism must perceive the relevant features of the environment in which it is to act; it must behave toward these objects in a

way relevant to the satisfaction of its impulse; and if all goes well, it then attains the phase of activity which is the consummation of the act. . . . A sign is *designative* insofar as it signifies *observable* properties of the environment or of the actor, it is *appraisive* insofar as it signifies the consummatory properties of some object or situation, and it is *prescriptive* insofar as it signifies how the object or situation is to be reacted to so as to satisfy the governing impulse.

At this juncture it is worth comparing the terminology used by Morris to describe the three aspects of signification with the terminology developed to handle the various components of psycholinguists and perception.[17] The designative aspects of signification may thus be regarded as essentially similar to what are called denotative meanings or the attributive component in perception.[18] Perception, it has been remarked, invariably involves an act of categorization.[19] A sign or stimulus is thus placed in a certain class by virtue of its defining attributes. When we ask what a sign designates, therefore, we ask what category it belongs to. This raises some fascinating problems regarding the interaction between perception and language since it is the latter which mainly determines the categories into which stimuli may be put. This is not the place, however, to debate the pros and cons of the Whorfian hypothesis that cognitive behavior is influenced by the semantic structure of language, although this hypothesis, if true, has important implications for the analysis of crosscultural differences in perception.[20] The *appraisive* aspect of signification may similarly be related to *connotative* meaning of the *affective* component of perception. A sign may provoke in us feelings of attraction or repulsion, like or dislike. From our point of view this is an important aspect of perception, because if a sign provokes a positive attraction then we can anticipate positive behavior with respect to it; if not we can anticipate negative behavior. If we can find some method of measuring the affective component of perception we will have gone some way to measuring a component of satisficing behavior. The *prescriptive* aspect of signification may also be related to the *expectancy* component of perception, although the relationship is not a perfect one by any means. A sign may provoke in us certain expectations on the basis of which we may make predictions. Each sign therefore stands in some relation to other signs and on the basis of these relationships we may make inferences of one sort or another. If, for example, the sign is the word combination 'rich suburb' then we generate certain expectations about the kinds of houses there, the people who live in them, and so on. If we wish to act in some way with respect to a 'rich suburb' we also generate certain expectations about what is a feasible course of action (a planner may find it to be the optimal location for a sewage processing plant but he may infer immediately that such a plant will be impossible in a 'rich suburb'). If we are to understand this component of the sign process, then we must understand the structure of associations which the individual possesses. Since these structures are relatively permanent it seems best to conceptualize them as attitudes rather than as perceptions. I suspect that it is at this point in the sign process that attitudes are most important and we cannot, therefore, ignore attitudes altogether in our analysis of the perceptual process.

Some signs, such as the word 'man,' are primarily designative (attributive or denotative); some, such as the word 'bad,' are primarily appraisive (connotative, affective); while others, such as the word 'should,' are primarily prescriptive. But all signs have some signification on all three dimensions.

How does this general notion of a sign process relate to the measurement of people's perceptions? The necessary operational concepts for the measurement of human perceptual behavior are provided by stimulus-response psychology.[21] The measurement of perception involves scaling (on some appropriate model) the responses of individuals to some stimulus. I do not wish to consider the problem of what constitutes an adequate scaling model, and I shall therefore take it for granted that the problem of matching the responses with some scaling system can readily be overcome. I want to concentrate upon the nature of the stimuli. All stimuli may be regarded as signs (although not all signs function as stimuli, e.g. certain logical signs such as 'and'). This elementary fact provides the link between the generality of the sign process and the particularity of measurement procedures in perception studies. The response which we scale is a disposition to act in a certain way as a result of an interpreter experiencing the sign. The problem of measurement in perception may thus be regarded as a question of how we select signs and how we can control the sign process so as to yield meaningful insights into spatial behavior.

It is useful at this stage to divide signs into *symbols* and *signals*. A symbol is a substitute sign, signifying what the sign for which it is a substitute signifies. If a sign is not a symbol we will call it a signal. Generally speaking our direct experience of an environment may be regarded as a sign process in which the signs are signals. The architectural form of a street, the morphology of a landscape, and so on, are mainly perceived through signals. Some elements of this experience may be conceptualized as symbols in certain contexts (e.g. the form of a church, or the layout of a village). We may, however, read about the same things in a book, in which case all the signs are symbols that are presumably chosen to represent the salient signals experienced in the environment. Now, the relationship between signals and symbols is an important one. In general our measures of perception in the social sphere are dependent upon symbols rather than signals. It is too expensive to take people to an area and monitor their responses during their stay there, and we therefore administer the symbol 'Devon' and measure their reactions to that. Most investigations into geographical perception are thus going to be through measuring reactions to symbols such as words, photographs, pictures, maps and so on. Yet actual behavior is often going to be determined by the perception of signals (shop-window displays, attractive views, and so on). In these situations we need, therefore, to calibrate the relationship between signal and symbol.

Let us consider the ways in which the signal sign process and the symbol sign process must necessarily differ. The context is obviously different. Symbols may be reacted to in a warm comfortable room, signals are experiences in a totally different environmental situation. The symbolic process can only indicate certain selected features of an object or generalize about it in such a way that a lot of information is lost (e.g. the relationship between the signals that emanate from Devon and the symbol 'Devon'). The signal process shows an object in its total environment. This is not necessarily a disadvantage for we have far greater control over the symbolic process as a stimulus. We can cut out background information on a photograph of a house, but we cannot take somebody to a house and ask them to look at it as if the background did not exist.

In defining a symbol it was suggested that it should signify the same thing as the signal it is designed to represent. Yet symbol and signal cannot be the same in every respect. We therefore need some way of defining the equivalence between signs. We cannot do this independent of a motivational situation.[22] The response to the symbol 'Devon,' for example, will depend upon whether we are contemplating taking a holiday or a permanent job. We can only talk of equivalence in the same motivational context. Equivalence cannot be established either without specifying the particular dimension of signification on which signal and symbol are to be equivalent. We might describe a suburb as 'nigger,' 'negro,' 'black' or 'colored'; on the designative dimension these symbols are not far apart, but on the appraisive dimension the symbols are very different. We may thus speak of signs which are designatively equivalent (they refer to the same object), appraisively equivalent (they arouse similar emotions of attraction and repulsion in us) and prescriptively equivalent (they suggest similar lines of action).

What of the symbols themselves? Since most human communication is through symbols and the learning process is largely an acquisition of symbolic tools, we must expect that symbolic representations are important in their own right and in many cases they attain a meaning which is independent of the signals which they were initially designed to represent. Consider the symbol 'Devon.' The relationship between it and the signals that emanate from that area of land called Devon is obscure. Suppose we want to predict decisions to migrate to Devon. Are these decisions made with respect to the signals that emanate from Devon or are they made with respect to the symbol 'Devon'? Symbols are often more important than signals in determining behavior, for the symbol may itself have an appraisive signification of some sort which has nothing to do with the signal process. Herein lies the difference between an individual's image of something (drawn from its symbolic representation) and the reality of that thing. Many an emigrant has been attracted to America, the land of opportunity, only to find that the image has been misleading in certain respects.

This raises an intriguing problem for perception studies in geography. We can, if we so wish, try to measure the designative, appraisive or prescriptive aspects of the symbols themselves. We may study people's reactions to words such as 'Devon' or 'Vermont' and try to establish the image attached to that symbol. We may also examine the perception of maps, photographs and so on in their own right. On the other hand, we may use the same symbols to try to study an individual's preferences with respect to a set of signals, such as the physical attractiveness of Devon or Vermont, or the signals represented by a map or photograph. Clearly, there is the possibility of confusion in the interpretation of measures obtained by using symbols as stimuli. One individual may interpret a symbol stimulus entirely at the symbolic level, another may do so with respect to the signals that a symbol represents, and some may have a mixed interpretation (partly symbolic and partly in terms of the signals). Which interpretation an individual gives to the symbolic stimulus will depend on a number of factors, but the one that is of obvious importance will be the amount of information and direct experience that a person has of the symbol and the signals that it represents. If a symbol such as a word or a map represents an area that I know well, then I am likely to associate the signals emanating from that area with the symbol, but if it refers to an area I have never had any experience of, then I will interpret it in symbolic terms. This symbolic interpretation will depend on the contexts in which I have experienced the symbol before. Suppose the only times in which I have come across the symbol 'Devon' are on travel posters that beckon me to holiday in beautiful Devon. Then it is likely that I will react favorably to the term on the appraisive dimension. If I have never come across a symbol before – say, 'Clackmannanshire' – then I have no word associations upon which to base my judgement. In this case I might simply judge the symbol on how nice it sounds.

Now it is quite usual in perception studies in geography to attempt to compare measures on different symbols over different people. This is clearly a fairly 'noisy' procedure if the above analysis is anything more than nitpicking. Let me give two examples. Suppose we ask people to rank counties or towns in order of their preference for living in them. We are here providing a list of symbols and measuring people's responses to them in a given motivational setting. In general we may expect that people have experience of towns or counties close to them (given the usual distance effect upon spatial behavior). We may therefore expect that the symbols of counties or towns close by will tend to be interpreted at the signal level. Further away, the symbolic interpretation is likely to be more important. I wonder if the tendency for people to react favorably to their home area and then to comment favorably upon certain other areas further away[23] can be explained by the mixing of a local signal effect with a more general symbolic effect? The same sort of comment can be made about the use of maps in geographical perception studies. If the map is used as a stimulus, do the results refer to the perception of the land represented by the map or to the perception of the map itself? Is it the country represented by the map that looks 'interesting' or is it the symbolic form that is 'interesting'? There are additional difficulties in using the map as a stimulus since it is a particular type of symbolic representation over which people have unequal command – some are good map readers and some cannot understand the relationship between the map and reality at all. Measures derived from maps as stimuli may, therefore, reflect the varying ability to think in terms of abstract spatial schema rather than measures of preference for those areas which the map represents.

In both of these examples we are really scaling several different things simultaneously: the ability to read the language in the particular symbolic form we have chosen, the response to the symbols themselves, and the response to the signals that the symbols represent. The variation in our measures may thus be explained by reference to several different sources of variation. It could be argued, of course, that variation from unwanted sources may be regarded as random noise, or that in any case it is the totality of the measure that is important if we are concerned with predicting geographical behavior. These arguments may be correct. But it seems important to have some empirical evidence, and this would not be too difficult to collect. We could, for example, compare a population group that has physical experience of Devon (and which presumably builds images with respect to the signals) and compare their images with those of population groups that lack such indirect experience. There are some interesting experimental possibilities here.[24]

The theory of signs, or semiotics, as it is usually called, provides a useful framework for the study of perception in geography. At the theoretical level we can conceptualize geographic behavior as the result of some sign process in which individuals are reacting to signals and symbols. Semiotics also has empirical relevance for it provides us with a framework for understanding the measurement problem and in particular the concepts of a signal and a symbol provide us with a means for differentiating between signs and establishing some sort of control over the stimuli used in measurement. The tridimensional concept of signification also has both theoretical and empirical relevance. It provides an underlying model for understanding the various facets of the perception process. We can, if we wish, examine sign processes on one of the dimensions only. Most attention has been paid to the appraisive dimension, and there are techniques for measuring the perform- ance of a sign on this basic dimension – the most important being the semantic differential which was specifically developed for the study of the connotative aspects of meaning.[25] Theoretically we may distinguish between reactions based on signal processes and behavior based on symbols. I suspect that we may find, for example, that local shopping behavior can best be understood as a signal process, whereas long-distance shopping to large centers may be conceptualized as a symbol process. Other aspects of behavior in the city may be affected by signals – Lynch, for example, has shown how distinctive architectural features in a city act as signals and, from this, we may expect that individual travel behavior may be analyzable by a study of the signals on various city routes.[26] On the other hand, processes such as emigration, migration over long distances, and so on, are more probably affected by the images attached to symbols. I suspect in many cases that signal and symbol components will be intermixed in various proportions.

There are many aspects of the sign process that I have not considered, of course. The human mind has a limited channel capacity and cannot accept all the signals that an environment sends out. The process of selection of signals is therefore important. We can think of signals of varying strength (a cathedral sends out a massive signal whereas an ordinary house in an ordinary suburban setting sends out a very mild one (unless it happens to be home)). We can measure the strength of various architectural symbols (Lynch's work is again interesting in this respect) and thereby make estimates of the probability that an individual will or will not perceive a particular signal. I have not space, however, to consider these various other aspects of perception in a geographical setting.

A cognitive–behavioral location theory?

I now want to draw upon the preceding discussion of theoretical and empirical problems to try to assess the prospects for formulating a cognitive–behavioral location theory.

It is inevitable in the early stages of any investigation that the concepts (both theoretical and empirical) that we use will be loosely defined. To demand exactness, precision and rigor in the use of terms in the initial stages of investigation can only result in 'the premature closure of our ideas' and hence have a pernicious effect upon the direction of research.[27] Yet we cannot afford to take this as a charter for interminable vagueness in our concepts. Indeed the degree to which we succeed in reducing this vagueness is a measure of our progress. But the reduction should be real and not spurious and forced. At the present time the concepts available to us for formulating a cognitive–behavioral location theory are rich in ambiguity, and I believe this to be necessarily so. I cannot avoid the suspicion, however, that the concepts current in geographical writing are unnecessarily vague. The concept of satisficing behavior is an excellent example. Similarly, we could do much more to pin down a conceptual apparatus for the study of perception in geography than we have done. The trouble here, of course, is that we must necessarily rely upon perception concepts as they are formulated in psychology and here we have considerable freedom of choice. Given our concern with geographical behavior, we are likely to find that much of the psychological literature on perception (and the conceptual apparatus contained therein) will be irrelevant to our purpose. But without a strong command of the psychological literature (which I for one do not possess) it is difficult to determine which presentations are useful to us and which are irrelevant. I am forced to conclude that we cannot hope to formulate a cognitive–behavioral location theory unless we breed geographers who have a strong command over the literature of behavioral science in general, or find behavioral scientists who are interested in geographical problems. The same sort of comment, however, can be made with respect to economic and stochastic location theory. In spite of the lip-service that geographers have paid to the former, there are few of us capable of adding to that theory mainly because we do not know enough economics. A stochastic location theory will similarly demand an adequate command over the mathematics of probability theory. But both economic and stochastic approaches to location theory have a distinctive advantage over the cognitive–behavioral approach because the concepts involved are far less ambiguous and much more easily pinned down. I therefore suspect that we may get further, more quickly, in developing economic and stochastic theory than we will in developing the cognitive–behavioral theory. If we are searching for immediate payoffs, therefore, we will do better to invest our time in furthering normative economic theory and in formulating stochastic theory; I believe the second possesses the greater untapped potential. But this has the unfortunate effect of leaving the domain of cognitive–behavioral events empty of any kind of formulation at all. I am sure that this is unacceptable and undesirable.

It is undesirable because the cognitive–behavioral domain contains problems that are real enough and significant enough – we surely cannot afford to ignore them. I suspect it is unacceptable because this domain is of enormous intrinsic interest. The problem of perception, for example, is so basic to everything we do and think and it is so basic to our understanding of knowledge itself. The cognitive–behavioral domain undoubtedly poses the greatest challenge of all. Stochastic theory, although of enormous potential, avoids so many intrinsically interesting problems. It may give us excellent predictive control over aggregate events but I fear it will never yield us a really deep understanding of process. I also take the view that research should not only be useful, it should be stimulating and fun. Here the cognitive–behavioral domain has distinct advantages!

We can enter this domain, however, with the expectation of obtaining only very limited results. I would doubt if anything very satisfactory will emerge in the way of general theory until the year 2000 AD or so. But certain small problems will, I think, prove tractable to theoretical analysis and empirical investigation. On the theoretical side, Isard and Dacey[28] have shown the way. On the empirical, some of the studies on the perception of resources and place preferences[29] have indicated that interesting and useful results can be obtained. If we refashion and sharpen our conceptual tools and improve our understanding of the measurement process, I have no doubt that substantially better results can be obtained. The one thing we cannot afford, however, is to indulge in that particular kind of intellectual laziness that regards it as unnecessary and foolish to try to eliminate vagueness and ambiguity in our conceptual apparatus. In this we must remain perpetually aware of the trade-off that exists between unnecessary ambiguity and premature rigor and adapt our research strategy accordingly.

Notes

1 See, for example, A. G. Wilson (1970) *Entropy in Urban and Regional Modelling*, London, Pion.
2 I have discussed the background to these strategies in D. Harvey (1970) 'Behavioural postulates and the construction of theory in human geography,' *Geographica Polonica*, 18.
3 E. Nagel (1961) *The Structure of Science*, New York, Harcourt, Brace & World; and F. S. C. Northrop (1947) *The Logic of the Sciences and the Humanities*, New York, Macmillan.
4 See R. B. Braithwaite (1960) *Scientific Explanation*, New York, Dover, 77.
5 H. M. Blalock and A. Blalock (1968) *Methodology in Social Research*, New York, McGraw-Hill, 5–27.
6 H. A. Simon (1957) *Models of Man*, New York, Wiley.
7 M. H. Segall, D. T. Campbell and M. J. Herskovits (1966) *The Influence of Culture on Visual Perception*, Indianapolis, Ind., Bobbs-Merrill, 24.
8 See, for example, K. E. Boulding (1956) *The Image*, Ann Arbor, University of Michigan Press; and D. Lowenthal (1961) 'Geography, experience and imagination: towards a geographical epistemology,' *Annals of the Association of American Geographers*, 51, 241–60.
9 W. Kirk (1951) 'Historical geography and the concept of the behavioral environment,' *Indian Geographic Journal*, Silver Jubilee Edition.
10 J. Wolpert (1965) 'Behavioral aspects of the decision to migrate,' *Papers of the Regional Science Association*, 15, 159–69.
11 See Harvey, 'Behavioral postulates' (see note 2 above).
12 P. B. Warr and C. Knapper (1968) *The Perception of People and Events*, New York, Wiley, 2.
13 ibid.
14 M. Fishbein (1967) 'Attitudes and prediction of behavior,' in M. Fishbein (ed.) *Readings in Attitude Theory and Measurement*, New York, Wiley, 483.
15 C. K. Ogden and I. A. Richards (1930) *The Meaning of Meaning*, New York, Harcourt, Brace & World, 78.
16 C. Morris (1964) *Signification and Significance*, Cambridge, Mass., MIT Press.
17 In psycholinguistics I am referring to C. E. Osgood, C. J. Suci and P. H. Tannenbaum (1957) *The Measurement of Meaning*, Urbana, University of Illinois Press; and R. Rommetveit (1968) *Words, Meanings and Messages*, New York, Academic Press; the perception material is summarized in Warr and Knapper, *The Perception of People and Events* (see note 12 above).
18 See Warr and Knapper, ibid., 7–13.
19 J. S. Bruner, J. J. Goodnow and G. A. Austin (1956) *A Study of Thinking*, New York, Wiley, 9.
20 See B. L. Whorf (1956) *Language, Thought and Reality*, Cambridge,

Mass., MIT Press; and Segall *et al.*, *The Influence of Culture on Visual Perception* (see note 7 above).

21 W. S. Torgerson (1958) *Theory and Methods of Scaling*, New York, Wiley.

22 D. E. Berlyne (1965) *Structure and Direction in Thinking*, New York, Wiley, 50–1.

23 P. R. Gould (1973) 'On mental maps,' in R. M. Downs and D. Stea (eds) *Image and Environment*, Chicago, Aldine.

24 I think the kind of research design developed in Segall *et al.*, *The Influence of Culture on Visual Perception* (see note 7 above) provides an interesting model, while J. Sonnenfeld (1967) 'Environmental perception and adaption level in the arctic,' in D. Lowenthal (ed.) 'Environmental Perception and Behavior,' University of Chicago, Department of Geography, Research Paper 109, provides a geographical example.

25 Osgood *et al.*, *The Measurement of Meaning* (see note 17 above); Warr and Knapper, *The Perception of People and Events* (see note 12 above); and R. M. Downs (1967) 'Approaches to, and problems in, the measurement of geographic space perception,' University of Bristol, Department of Geography Seminar Papers, A, 9.

26 K. Lynch (1960) *The Image of the City*, Cambridge, Mass., MIT Press.

27 A. Kaplan (1964) *The Conduct of Inquiry*, San Franciso, Ca., Chandler, 62–71.

28 W. Isard and M. F. Dacey (1962) 'On the projection of individual behavior in regional analysis,' *Journal of Regional Science*, 4, 1–32 and 51–83.

29 For example, R. W. Kates (1962) 'Hazard and choice perception in flood plain management,' University of Chicago, Department of Geography, Research Paper 78; T. F. Saarinen (1966) 'Perception of drought hazard on the great plains,' University of Chicago, Department of Geography, Research Paper 106; Gould, 'On mental maps' (see note 23 above); and P. R. Gould and R. White (1968) 'The mental maps of British school leavers,' *Regional Studies*, 2, 161–82.

3 The geographical relevance of some learning theories

Reginald G. Golledge*

The search for explanations of the spatial behavior of individuals and groups inevitably leads to a discussion of the processes that influence behavior. Recent emphasis in geography on interaction, diffusion and decision-making models, and a surge of interest in some spatial aspects of psychophysical theories of perception, confirm this trend. Another process that involves some useful spatial concepts, but which so far has merited scant attention in geography, is the learning process. It is the aim of this paper to examine the role of this process in spatial behavior, to indicate some useful spatial concepts from learning theory, to review a selection of learning models that could conceivably be used in a spatial framework and to suggest some problems that are suitable for analysis by the models presented in the paper.

Learning theory and spatial behavior

For the purpose of this paper 'spatial behavior' is defined as any sequence of consciously or subconsciously directed life processes that result in changes of location through time.[1] 'Learning' is the

* This is an abbreviated version of the 1969 paper; sections on the mathematical representation of concept identification models, paired associate models, stimulus ranking models, Luce's choice theory, and various Markov models have been eliminated from this draft.

process by which an activity originates, or is changed, through responding to a situation – provided that the changes cannot be attributed wholly to maturation or to a temporary state of an organism.[2]

Geographers have concerned themselves with a variety of spatial behaviors. For example there exists a large volume of cross-sectional studies of consumer,[3] journey-to-work,[4] production behavior[5] and so on. Such studies examine movement patterns through space, generally over a very limited time period.

Perhaps the most frequently occurring reference to spatial behavior in geographical literature occurs through the use of 'spatially rational man' assumptions. Just as classical economic theory postulated the existence of economically rational man, so geographers have incorporated into their theories postulates that assume that man may have perfect knowledge of alternative trip types, a stable system of ordered space preferences for goods and services, and a constant preference for the least effort solution in trip making.[6] It is eminently possible to theorize about such a man, and he has appeared consistently in geographic literature.

Recent papers have stressed that spatial rationality is but one of a range of behaviors that actually occur in space. For example, Wolpert[7] has examined the notion of bounded rationality or satisficing behavior; Gould[8] has examined search behavior; Golledge and Brown[9] have shown that both search and stereotyped behavior can be observed at different time periods in the marketing act; Huff[10] has attempted to construct the motivational basis of spatial behavior; Marble[11] has investigated asymptotic behavior in journey-to-work patterns; and Pred[12] has tried to construct a behavior matrix depicting the dynamics of locational decision making. Techniques used in each study vary from choice models to Markov chains, renewal processes and topology.

Problem solving and habitual behaviors

Each of these cases cited above show that the type of behavior with which geographers have most frequently concerned themselves is problem-solving behavior. They have examined such things as the decision processes that result in locational choice: the problem of locating urban functions; the problem of choosing paths to work, to shop and to play; and so on. Once the problem appears to be solved, consequent behavior is neglected. It is pertinent at this state briefly to examine problem-solving behavior and its relation to other spatial behaviors.

Problem-solving behavior is noticeably different from weakly motivated (or random) behavior and is also different from habitual response making. It is characterized by:

a confrontation with a problem;
b deliberate thinking in a specified direction (search, or vicarious trial-and-error behavior); and
c choosing among alternative courses of action and making overt responses.

Problem-solving behavior is a highly selective process, and represents the attempts of mankind to adapt to changing conditions. Changes in behavior due to problem solving are often substantial and abrupt. Despite the apparent importance of problem-solving behavior, however, it has been claimed that it is not the most common form of human behavior.[13] Routine behavior or habitual-response behavior is claimed to be more frequent. Routine behavior generally follows problem-solving behavior and is of greater duration through time. However, if environmental changes are frequent, or if several alternatives of approximately equal magnitude or reinforcement are faced, behavioral oscillation[14] may occur. The effect of this acitivy on a model's predicting ability is obvious. Routine behavior on the other hand often follows a path of minimum effort, it serves to reduce uncertainty in the decision process, and reduces consideration of alternative courses of action. In other words it is the behavior most used to cope with the contingencies of everyday living.

Learning and behavior

Regardless of the type of spatial behavior that is observed at any time or that evolves through time, it is apparent that such behavior is a learned phenomenon. Consequently, some knowledge of the learning process is essential to understand it. For example, the degree of correspondence between theoretical and observed behavior can be interpreted as a function of the extent of complete learning in a given system. It is feasible to assume that at any given time, and for any existing (non-controlled) population, we can expect part of a population to exhibit forms of habitual behavior which might be explained simply by least effort or other specified characteristics. However, we can also expect a portion of the population to be in earlier phases of learning about the system in which they live. Recent immigrants will exhibit some type of spatially irregular search behavior; those partly acquainted with the system in which their behavior is observed may have partly formalized behavior; those with considerable experience of the system may exhibit forms of habitual behavior. We can imagine that all members of the system are striving toward some asymptotic spatial response. Theoretically, this may be complete spatial rationality; or it may be a series of habitual responses that represent a satisfactory coping strategy, but which are not limited to selection of a single alternative. Approach towards this asymptote varies with the accumulation of knowledge about any spatial system and extent of experience with it.

Spatially relevant learning concepts

Many concepts from learning theory have direct spatial application. By examining briefly a selection of theories, it is possible to isolate some concepts that appear useful for geographers.

Action space

Recently geographers have found considerable value in the concepts of 'life space,' 'action space' or 'ego space.'[15] Extensive use of these concepts was made by Kurt Lewin, who regarded behaving organisms as geometrical points moving about in life space.[16] Individuals were subject to the pushes and pulls of personal and group expectations. In the course of their locomotions through space, individuals circumvented barriers of one sort or another, and their locations at any point in time were determined by the forces impinging on their life space at that time. By adding the characteristic of mobility to each individual in his life space, Lewin envisaged movement towards locations that were adient (or which had a positive valence), and moving away from vectors in the life space that were abient (or which had a negative valence).[17] In geography, ideas similar to this appear in migration theory (e.g. the push–pull hypothesis) and the concept is implicit in the idea of town attractiveness, and the centrifugal–centripetal forces literature in urban geography.

The presence of adient–abient vectors in life space gives rise to conflict situations. Examples of these are:

a where two positive valences of equal force exist (leading to an approach–approach conflict similar to the concepts implied in the selection of market centers);

b where two negative valences exist (leading to avoidance–avoidance situations); or

c where there are simultaneously present positive and negative valences (leading to the attraction–friction combination used frequently in geography).

Lewin also argued that only where life space (which is perceptual) and physical space (which is actual) coincide, can Euclidean-space distances and directions be used to describe either the locomotions (paths) or the responses (movements) of people. Otherwise he argued that some type of space transformation should be used. It is interesting to note that Tobler, Getis, Gould and other researchers are presently experimenting with hodological-space measures of the concepts of distance and direction for possible use in geographic models.[18]

The critical feature of Lewin's work is, however, that he argued that accumulated information about a system in which each individual operates is a key factor influencing both locomotions in space, and the gap between perceived and physical space. In other words, the ability of an organism to cope with a space system depends on what he can learn about it. Obviously this statement has as much relevance for the geographer interested in behavior as it does for the psychologist interested in the learning mechanism.

Place learning

Tolman's learning theory argues that an organism learns not by learning movement habits, but by learning the location of paths or places. In other words, he argues that learning is a cognitive process guided by spatial relationships rather than by reinforced movement sequences.[19] This theory is known as sign learning and it can be defined as an 'acquired expectation that one stimulus will be followed by another provided that a familiar behavior route is followed.'[20] The movement that results is variable; i.e. one movement may be substituted for another provided that both movements lead to the same end point where a stimulus or reward is expected. Habit-formation theories (conditioning) assume that what the organism learns are movements or responses to stimuli. Sign learning proposes that under some circumstances the organism, instead of learning movement habits, learns the location of paths and places!

Latent learning also supports the theory that spatial orientation rather than sets of movements are learned. Latent learning refers to any learning that is not demonstrated by behavior at the time of learning.[21] Such learning occurs under low levels of drive, or if drive is inactive, when incentives to learn are lacking. When drive is heightened or incentives appear, there is recall and use of what was learned. For example, through 'mild curiosity' we note the location of a store selling goods in which we are not at the moment interested. However, when we want something the store sells, we head directly for it even though it was never before a goal object.

In association with this notion that mental maps influence behavior, Tolman also investigated the problem of 'reward expectancy' or levels of aspiration. When a shopping trip for (say) food is undertaken, then obviously there are expectations concerning the type and quality of food that is sought. If a reward is merely something that satisfies a drive, one type of reward (e.g. food, clothing) ought to be as useful as another type (i.e. food, clothing) as long as it is generally regarded as 'acceptable.' But if a certain kind of reward is expected, alternatives that were not expected may be rejected. For example, many people shop for specific brands of goods. If at a given store the expected brand is not available then the consumer may go elsewhere – even though an adequate array of generally acceptable substitute brands are available! Thus, multiple-place shopping may occur when a given level of aspiration is not achieved, even though close substitutes for an item or service are available. This concept has been implicitly used in geography to account for unexplained variation in shopping-behavior models.

Contiguity

Geographers have traditionally explained the location and distribution of phenomena by looking for variables areally associated with or contiguous to the problem variable. Guthrie's learning theory should be inherently appealing to geographers because it argues similarly that the influence of a rewarding state of affairs acts on not only a given stimulus-response connection, but also on acts in the neighborhood of a rewarded one.[22] Thus there is a spread effect in learning which reinforces responses in the vicinity of (or which are associated with) rewarded response.

Guthrie's theory of learning is also based on the conviction that behavior can be studied only when it is overtly observable. Very frequently overt acts have some spatial manifestation – and these spatially overt acts are precisely the behavior that geographers try to explain. This is then a theory concerned with movements themselves rather than whether movements lead to success or error.

One further aspect of this theory that has led to its mathematical interpretation, is the argument that 'the animal learns to escape with its first escape. . . .'[23] This has led to the interpretation of learning as a Bernoulli-type process and ultimately to the representation of learning as a Markov process.[24] He also called attention to repetition and stereotypy in behavior after learning was achieved, and, of course, this is the type of behavior geographers frequently seek to describe and predict. Another interesting conclusion from his work was that rewards do not strengthen behavior but only prevent it from disintegrating. Although little research into extinction has been undertaken in geography, it also promises to be a fertile field of inquiry.

Habit

The central concept of Hull's theory is habit.[25] Most of his information about habit formation is derived from experiments with conditioned responses (reinforcement theory). A major thesis of this theory is that, rather than contribute its maximum influence on one trial, reinforcement adds an increment to habit strength on each occurrence. Habit strength is defined as a positive growth function of the number of trials.[26] In addition to conditioning, Hull examined such phenomena as trial-and-error behavior, discriminatory learning, maze learning, and other useful concepts. The end result of this research was an incremental learning theory.

The theory that learning is a gradual process includes within it the notion that behavior changes from motivated search (trial and error) to fully learned activities. Since this transition involves a variety of spatial manifestations, it is pertinent to examine in detail some of its concepts. For example, many problems require the selection of one or another mode of action to reach a goal. Often alternative responses appear successively, in an unorderly fashion, until a satisfactory response is made. Such procedures are called 'trial-and-error procedures,' and their basic element is search activity. If an individual is placed in an unfamiliar environment and stimulated to seek a goal, he generally exhibits a tendency to vary his responses (the 'law of multiple responses'). These responses are made under conditions of uncertainty as to their outcome. After a range of trials, the 'correct' or most satisfying response is retained (the 'law of effect').[27]

In search practices, prior experiences will influence the type of search patterns; this is particularly true if there is a transfer of knowledge from previously experienced search systems. The first stage of a search procedure is called the 'provisional try.' This is an attempt to find a satisfactory response pattern and is sometimes referred to as hypothesis behavior. The provisional try achieves an outcome and is corrected by the feedback of information concerning the consequences of the try. Thus, a try that is rewarded is favorably reinforced and may become the basis for the formation of an habitual response. A try that is unsuccessful or reinforced by punishment may be deleted from the response pattern and an alternative search pattern substituted for it. Geographically we recognize some regularities in the search process. For example, we often adopt an assumption that search will take the form of trying first the closest alternatives and trying last the furthest alternatives (least-effort hypothesis). Alternatively, we often argue that there is a definite trade-off between expected satisfaction from a trip and distance, such that in multiple-response situations there develop regular 'decay functions,' which specify probabilities that search will be extended over certain distances. In fact, both these assumptions form the basis of our most powerful urban theories.

The principal idea to be retained at this stage is that, while search may be a random procedure, it is more likely to be undertaken according to a definite set of rules. In fact, it can be argued that once beyond the first sensorimotor experiments, an individual's responses will never be purely random. We can even postulate at this stage that stereotyped search procedures may develop in which outcomes are sampled systematically according to, say, a locational or directional bias. The degree of success resulting from this type of searching may govern the degree to which an approach is made toward a rational or maximizing or habitual behavioral pattern.

Hull also recognized that there are multiple-response paths between any origin and goal, and consequently individuals learn alternative ways of traversing routes. These alternatives form a habit 'family,' which are arranged in a preferred order.[28] For example, short routes may be more strongly preferred to long routes. Less-favored routes are chosen when the more favored are blocked in some way. Types of spatial barriers and their effect on route selection have been defined by Yuill[29] and by researchers interested in diffusion models.[30] The habit-family concept is useful for examining the problem of the attraction of goals behind barriers and the selection of paths around barriers. It is worthwhile noting here that many geographical models of spatial behavior have used an assumption of free space (no barriers), and often modifications of theory involve introducing barriers and examining resulting changes in movement systems.

The concept of stereotyped responses

The principal characteristics of a stereotyped response are rigidity (invariability), repetition, and resistance to change (persistence). Stereotyped actions consist of constantly elicited patterned responses that have a high degree of habit strength.[31]

Most geographical theories that include behavioral elements and produce optimizing predictions rely implicitly on the assumption of stereotyped behavior. In many cases the 'stereotyped' assumption is necessary in order to have any theory at all. Traditional city-hinterland theories for example assume the existence of a least-effort syndrome that produces a stereotyped action of always patronizing the closest occurrence of a phenomenon. Unfortunately, the stereotyped action assumed in this case is a trivial one – the repeated patronage of a single node. In practice, while recognizing that stereotyped actions do occur, most of the actions involved in various types of marketing behavior are of the multiple-response type. This is true of both shopping trips within a city and patronage of urban centers by the dispersed population.

One of the most significant factors related to the existence of spatial stereotyped action is that its occurrence makes possible the formulation of spatial behavioral theories. We can in fact regard stereotyped responses as asymptotes of a learning process; once this stage is reached then it becomes a simple matter to predict future responses. However, we do not know how long it takes an individual or a group to achieve a stereotyped phase, nor do we know what proportion of any given group at any time has already achieved stereotypy and how many are still in a search phase. This means that the degree of successful prediction by models which include stereotyped action as a necessary condition may be very small. In fact, this is the situation experienced when attempts are made to apply existing geographic models of consumer behavior to real-world conditions and deduce from them the locational patterns of some of man's activities. Despite insufficient empirical findings about stereotypy, we can suggest as a hypothesis, for the time being, that stereotyped actions in space develop as part of a process that aids the organism in adapting to its environment by providing it with a relatively organized and systematized conception of the multiple and everchanging events and experiences of life. Without such a conceptualization, it would be very difficult to find any regularity in the actions and behavior of humans.

Choice and the decision process

An essential element of the learning process is the choice of responses. Some theories describe learning as a change in response probability as the result of correct and incorrect choices in a continuing decision process.[32] Within the field of geography, Huff, Thompson and Wolpert have discussed various aspects of the decision process and corresponding choice behavior.[33] For example, Huff has attempted to conceptualize the transition from a premotivated stage of behavior to the overt spatial act. This disregards quiescent or unmotivated behavior (such as neural or synaptic responses), which are of little concern to the geographer, but it also neglects the recursiveness associated with the assessment of an overt act due to levels of aspiration and the activities of learning. In short, once an overt spatial response has been made, individuals will probably restructure their decision processes in the light of accumulated information (i.e. after the basic unit of learning has been achieved). Either the provisional-try behavior will be repeated or a new response choice will be made. Eventually a firmly established response pattern will emerge, which is regularly triggered by presentation of the original stimulus condition. The asymptotic result of a repetitive choice process is the formation of stable-state choice proportions, or strategies, that govern future behavior.

The examination of choice processes also provides information on the idea of 'rational behavior.' In the past geographers and other researchers have considered 'rational' and 'economic maximization' to be synonymous. Siegel, on the other hand, shows that both pure and mixed (matchings) strategies can be considered 'rational.'[34] He equates economic rationality with the development of pure strategies, but also argues that if boredom, curiosity and other motivations are considered, economic gain may not dominate behavior strategies. In fact he constructs a model of the choice process with two critical choice components: the utility of correct choice and the utility of variability. It would appear that geographers can make considerable use of both concepts in their own research on choice behavior.

It is apparent from the foregoing discussion that a number of learning theories and many of their key concepts have definite spatial implications. The possibility of using such theories in geography is enhanced by their translation into objective mathematical models which can be applied both to spatial and nonspatial situations. The following sections provide a review of some recent learning models together with suggestions for their spatial applications.

Recent learning models with possible spatial applications[35]

The recent attention to the representation of learning through the media of mathematical models has led to a refinement and operationalizing of many elements of learning theories. Considerable attention in current research is being focused on:

a association models, e.g. concept-identification, paired-associate, and linear-operator models;
b stimulus-sampling models, based either on linear-difference equations or stochastic processes;
c interactance-process models, based on either game theory or Markov models; and
d avoidance-conditioning models using linear-difference equations.

Most current learning models are probabilistic in nature. This implies that some variability of behavior is built into the models. By interpreting behavior as a stochastic process, for example, it is argued either that it is intrinsically probabilistic, or that it is determined by antecedent conditions that encourage variability from one time period to the next.

Geography is sometimes characterized as being in a phase of development in which it is still trying to discover how to measure its major dependent variables. To some extent we can claim affinity with psychologists such as Guthrie, Skinner and Estes, who have argued that response probability is their most appropriate dependent variable. Certainly there are many spatially overt acts which we measure in terms of 'response probability' – i.e. frequency of contact, etc. – and which we can classify into different types of response classes. Thus, models that have spatial variables as inputs, and which illustrate changes in response probabilities over time, appear suitable for both geographic and psychological uses.

The concept-identification model

Perhaps the most elementary learning model that has potential use for geographers is the 'concept-identification' or 'discrimination-learning' model. This is presented either from the stimulus-response reinforcement viewpoint of Hull, Spence, Burke and Estes,[36] or as a model of hypothesis behavior, as in the work of Restle or Levine.[37] The former viewpoint argues that subjects discriminate among responses according to type, force and frequency of reinforcement and that the end product is habit formation. The hypothesis viewpoint argues that responses are elicited by a set of stimuli after trial-and-error behavior has occurred. Such trial-and-error behavior is essentially a sequential testing of hypothetical responses, with the aim of finding a satisfactory stimulus-response pairing. The discrimination-learning model is based on the task of finding which cue is relevant in a given problem situation. Once the cue is recognized, future responses are based on this. In other words, the individual has learned a response pattern and may develop a habitual response to it. Market researchers have used these concepts to show the value of labels and advertizing in the selection of goods.[38] Geographers can also examine analogous conditions. For example, in the development of a journey-to-work pattern, we presume that travelers select alternative routes from a total array of routes and then test them. If a choice (cue identification) leads to an error or unsatisfactory trip sequence, that route may be discarded and a new choice made. Cues that influence selection of routes have so far been identified as least effort, shortest time, maximal aesthetics, fewest barriers, least boring route, and so on.

Despite the large number of criteria that may influence choice of a response path, generally there are cues that are more obvious than others. For example, for high-order (shopping) goods the most relevant cue for achieving a satisfactory response might be one of

Table 3.1   *Cues for selection of shopping centers*

Attribute	*Frequency ranked no. 1*	*Proportion of times ranked no. 1*
Closeness	112	35.7
Variety of stores	90	28.7
Quality of products	32	10.2
Parking	28	8.9
Prices	15	4.8
Service quality	6	2.0
Freeway access	4	1.7

Source: 'Shopping Center Utility Survey' (1967) Department of Geography, Ohio State University

the following: the presence of a department store in the center chosen for shopping; a minimum number of functions in the chosen center; or the distance to be traveled. Irrelevant cues might be: whether or not a friend recommends the center; presence of a cafeteria; presence of a movie theater; or presence of more than twenty-five functions.

Once a subject learns which cue is most relevant, he bases future actions on this. For example, Berry has argued that the number of functions is a relevant cue for predicting probabilities of trips to urban centers.[39] Golledge, Rushton and Clark argue that a maximum distance is the most relevant cue.[40] A study conducted in Columbus, Ohio by the author resulted in identification and ranking of cues in the order given in Table 3.1.[41] Support for both the above hypotheses is evidenced by the data collected in this study. One conclusion drawn from this study was that, as we alter the trip purpose, the relevant cues change, but once a relevant cue has been identified for a specific purpose, it influences future behavior.

Paired-associate models

Paired-associate learning models appear to hold considerable promise for the geographer. In this case there is probably equally as much potential in the way data are collected and prepared as there is in the models using such data. There are a number of conventional models used in paired-associate experiments, including one-element models,[42] single-operator linear models[43] and random-trial increment models.[44] For purposes of convenience only the single-operator linear model will be examined here.

The single-operator linear model views learning as a direct change in response probability from one trial to the next.[45] Since this type of model has already been investigated by geographers, no examples will be developed here.[46] However it is relevant to suggest some spatial characteristics of the model. The literature of geography abounds with market-area studies. One of the main conclusions drawn from these studies is that, as distance from a given center increases, the likelihood of patronizing it diminishes. In other words, decreasing probabilities of patronage can be partly explained by a decrease in the probability of learning about the advantages of a center as distance increases. Alternatively it can be argued that, as distance changes, the likelihood of approaching an asymptotic pattern limited to single-center patronage is correspondingly diminished. In terms of the linear-operator learning model, both the λ_i (asymptotic values) and α_i (learning parameters) may vary over distance. At present no evidence exists to show the manner of this variation, but the calculation of this change through space presents a challenge to the geographer.

It can also be expected that statistics derived from the model will vary consistently with distance, so that mean total 'errors' will increase as will the number of 'trials' before any asymptote is reached. The relative advantage of this information for different classes of goods (and therefore different sizes of center) is obvious and will not be further developed.

Interactance-process models

One of the most widely occurring problems in the social and behavioral sciences is the problem of predicting the outcome of a choice process when two or more alternatives are placed in competition with one another. A variety of models have been derived to handle choice decisions, ranging from locational-choice models in geography,[47] to resource-allocation models in economics,[48] and preference-ordering models in psychology.[49] The complexity of such models varies from simple gravity models used to predict the choice of market centers, to complex gaming models with mixed strategies. The versatility of many choice models allows them to be used both to describe the choice process and to describe the resulting interactions that occur.

Descriptions of the choice act generally specify the alternative eventually chosen (*a posteriori* choice), the nature of any trial-and-error behavior that occurs, and the time taken to respond, as well as including an expression of confidence in the correctness of choice, and a statement of the difficulty experienced in making choice.

Geographers have most frequently dealt with *a posteriori* choice and have neglected other aspects of the process. This is largely because overt spatial activity results from the choice act, and it is this activity that most interests the geographer. The data recorded by geographers are simply the alternatives selected and then spatial and other variables are used to explain why the specific alternatives were chosen. Sometimes possible response strength (or 'potential') is estimated from a series of overt spatial acts, and potential-type models are used to predict future interactions. It is interesting to note that recent geographic research has gone beyond analysis of *a posteriori* choice to examination of the antecedent and concurrent conditions influencing choice.[50] It is for this type of research that variations of learning models become relevant, and assumptions such as the weak stochastic-transitivity assumption, Luce's choice axiom, the constant-ratio rule, and the scaling theories of Togerson and Coombs become significant.[51] Researchers contributing models of this type include Hull, Spence, Simon, Atkinson, Suppes, Kruskall, Shepard, Carroll and numerous others.[52]

Luce's choice axiom concerns itself with the relationship between choice probabilities as the number of alternatives involved in the choice act change.[53] Its basic premise is that the ratio of the likelihood of choosing element a to the likelihood of choosing element b in the set of k-alternatives is a constant irrespective of the number and composition of the other alternatives in the choice set. For example, a sample of potential or actual shopping-center patrons are presented with a list of three possible centers and, using paired-comparison procedures, are asked for their choices between the centers. Results may be as shown in Table 3.2.

Table 3.2   *Paired comparison of shopping centers*

Shopping center	Paired comparisons		
	1	2	3
1	—	0.60	0.90
2	0.40	—	0.80
3	0.10	0.20	—

Source: Hypothetical data

If we were given results for $\Pr(1,2) = 0.60$ and $\Pr(2,3) = 0.80$, we could estimate $\Pr(1,3)$ from the choice axiom as follows:

$$w_{12} \cdot w_{23} = \frac{\Pr(1,2)}{\Pr(2,1)} \cdot \frac{\Pr(2,3)}{\Pr(3,2)}$$

$$= \frac{0.60}{0.40} \cdot \frac{0.80}{0.20} = 6 = w_{13}$$

and

$$\Pr(1,3) = \frac{w_{13}}{1 + w_{13}} = \frac{6}{7} = 0.86.$$

Despite the apparent promise of this choice model, empirical testing of its dominant thesis in a spatial framework has not yet produced good results.[54] Apparently the reason for this is that it is useful for repetitive-choice situations for individuals, but cannot easily be aggregated to reach conclusions for group behavior, unless it is assumed that all individuals in the group had the same basis for stating preferences. With a population of individuals whose locations vary, this limiting assumption is not feasible.

A more general use of Luce's work can be found in the realm of stimulus ranking. In ranking experiments that are not conducted by pairwise-choice procedures, stimulus-ranking theories, using the choice axiom, allow us to predict the probability with which each rank occurs. In this case, the rate of change in response strengths from trial to trial is interpreted as the learning parameter.

The stimulus-ranking model also involves calculating probabilities for ranking experiments. Ranking of a set of k-alternatives amounts to advancing a hypothesis about the way people generate and order their preferences. It is possible to use stimulus-ranking theory to predict the probability with which each ranking will occur, and the relative order of the rankings. By itself this amounts only to a cross-sectional description of the order of choices. However, when combined with choice theory and placed in a stochastic framework, it is possible to account for features of choice behavior in addition to specifying the alternative chosen; for example, the trial-and-error behavior prior to final decision making, the time taken to arrive at a fixed choice pattern, the degree of confidence in the decision, and so on, can all be included in the one model. One interesting conclusion that can be drawn from this type of model is that, when stimuli are of low attractiveness, there will be a large amount of trial-and-error behavior before a state of 'absorption' is reached. This means that, by examining the behavior of individuals with respect to selection of alternative paths, or alternative goals, we can estimate the relative strength of the goals. This obviously has some interesting ramifications for studies of the dynamics of market-area analysis, and could perhaps be used to help explain results from gravity models, Reilly's law, and other devices aimed at predicting patronage of centers.

Avoidance-conditioning models

Avoidance-conditioning models can be represented either by the two-element linear model of Bush and Mosteller or the nonlinear β-models of Luce.[55] Both these models consider that learning takes place on each trial whether the outcome is reward (avoidance) or nonreward (shock). They also argue that reward and nonreward trials do not have equal effects and they formulate models to account for these.

The general form of the linear model is:

$$p_{n+1} = \begin{cases} \alpha_a p_n + (1 - \alpha_a)\lambda_a \text{ if shock is avoided on trial } n \\ \alpha_s p_n + (1 - \alpha_s)\lambda_b \text{ if shock occurs on trial } n, \end{cases}$$

where

p_{n+1} = probability of successfully completing the $(n + 1)$st trial (i.e. avoiding shock),

p_n = probability of avoidance on trial n,

α_a = a fraction representing the extent to which success probability is increased on each trial when avoidance occurs,

α_s = fraction representing the extent to which success probability is increased on each trial when shock occurs,

λ_a = an asymptotic probability that could be reached as avoidance continues and

λ_b = an asymptotic probability that could be reached if shock continues.

In terms of the learning parameter (α), the bounds are $0 \leq \alpha \leq 1$. In this case the point where $(\alpha = 1)$ corresponds to the case where no learning occurs and $(\alpha = 0)$ corresponds to complete learning. Thus the smaller α is, the faster learning occurs. Values of α_i are usually estimated from tables provided by Bush and Mosteller.

  This model also has considerable potential for geographers and has been used in one-element form by Golledge and Brown, Haines, and Burnett.[56]

Spatial problems and learning models

Learning theories and their mathematical models describe changes in behavior from the first motivated act to a fully learned response or response sequence. The output from learning models can be used to describe all forms of spatial behavior from initial searching to repetitive (habitual) behavior. The entire learning sequence can be described by such models, or, alternatively, behavior at any point in time can be represented.

  The versatility of learning models makes them potentially useful for geographers because the spatial actions with which we deal invariably encompass a range of behaviors from search to stereo-typed habitual responses.

  Frequently we have criticized optimal models of spatial behavior because they apparently do not fit 'reality.' Evidence for rejection is inevitably obtained by comparing optimal behavior with some empirical evidence of actions. Just as inevitably this empirical evidence is a conglomerate, composed of the actions of an unstratified group of people who may be at different stages of the learning process. Thus we have compared theoretical asymptotic behavior with a mix of search, partly learned and fully learned behaviors, and as a result achieve a low level of explanation – which leads to rejection of the model, reinterpretation of theory, and perhaps a frustrated researcher.

Obviously one way of overcoming this problem is to test models with cross-sectional data that are relevant and appropriate. An alternative is to undertake further research into behavioral processes and thus more fully understand the complex 'reality' we often (rather naively) try to predict.

As pointed out by Simon, Katona, Wolpert, Gould, Golledge and Brown, and other interested researchers, few repetitive-choice decisions are made without some preliminary search activity. One of the problems begging further research is the nature of space-searching activity. Gould[57] has offered a fine introduction to the problem, and I believe considerable progress could be made by using learning models.

However, not all spatial activity is search activity. Some regular or habitual behavior pattern gradually evolves as a coping strategy. Of course, as this type of behavior evolves, the amount of space-searching should diminish. The formation of habitual response patterns involve stabilizing movements in space, for this means that such movements are predictable with a high degree of accuracy. The development of habitual behaviors is important for studying journey-to-work, shopping and recreational behavior – in fact, a variety of social and economic interactions where repetitiveness is involved. Thus, examination of the behavior of stratified groups of immigrants and of long-time residents would precede the building of predictive models, and again the format of some learning models appears appropriate for such an examination.

One of the consequences of migration within urban areas is the extinction of some previously held responses and acquisition of new spatial habits. Both these problems can be studied via learning models. However, the acquisition of new spatial habits is not limited to recent migrants. Each time a new urban function or a new response path develops, there is some effect on existing behavior. Impact studies of this type can be examined by both stimulus-sampling and signal-detection models, and the output from such models can be used to simulate future urban behavior.

Other problems that could benefit from study within a learning-model framework include: examining the role of pretrial information on trial behavior; finding the effect of system 'shocks' such as new avenues of transportation or new 'rewards' on spatial behavior; experimenting with extinction processes as various types of spatial barriers intervene between origins and destinations; and seeing if extinction of a response in one segment of space inhibits interaction with spatially associated goals.

Probably one of the most interesting and potentially useful outgrowths of experimentation with learning models comes not so much from a particular model, but more from manipulations with pairwise choice and ranked data. This leads the researcher to the field of metric and nonmetric multidimensional scaling models, some of which are reviewed in Rushton's paper in this volume. Use of such models would hopefully allow examination of subjective utility or preference rankings and should be particularly important to the field of consumer behavior.[58]

This list of models and problems is not comprehensive; it only scratches the surface of a range of spatial problems that could have some light thrown on them by using ideas and models developed by learning theorists. The least impact of such a move would be to encourage more frequent use of the dynamic approach to behavioral problems, and help geographers move away from cross-sectional analysis and the limited-use static models that have served so well in the past.

Currently research is being undertaken on a variety of spatial–behavioral problems. Interactance-process models (i.e. choice models) are frequently being used,[59] and attempts have been made to use linear-operator models (in market-decision studies)[60] and some forms of signal detection model (in diffusion studies),[61] while the literature on Luce's choice model, stimulus-ranking models, and multidimensional scaling has grown immensely. It is hoped that continued emphasis on spatial behavioral problems will lead to use of some of the other type of learning models summarized in this paper.

Notes

1 Such a definition does not limit behavior to humans, but it does eliminate the actions of inanimate objects such as rocks rolling, winds

blowing, plants growing, etc. It *does* allow us to infer that behavior is *caused* and has *directness, motivation, action* and *achievement.*

2 This process involves a number of identifiable steps from making the first motivated response, through a series of covert and overt behaviors until a basic unit of learning has been achieved. For a summary of definitions of learning see E. Hilgard (1956) *Theories of Learning* (Second edition), New York, Appleton-Century-Crofts, 2–6.

3 B. J. L. Berry, H. Barnum and R. J. Tennant (1962) 'Retail location and consumer behavior,' *Papers of the Regional Science Association,* 8, 65–106; S. H. Britt (1966) *Consumer Behavior and Behavioral Science,* New York, Wiley; W. Isard and M. Dacey (1962) 'On the projection of individual behavior in regional analysis,' *Journal of Regional Science,* 4, 1 and 2, 1–35 and 51–84; E. N. Thomas, R. N. Mitchell and D. A. Blome (1962) 'The spatial behavior of a dispersed non-farm population,' *Papers of the Regional Science Association,* 10, 107–33; G. Rushton, R. G. Golledge and W. A. V. Clark (1967) 'Formulation and test of a normative model for the spatial allocation of grocery expenditures by a dispersed population,' *Annals of the Association of American Geographers,* 57, 2, 389–400.

4 References on this topic are summarized in B. J. L. Berry and A. Pred (1965) *Central Place Studies,* Philadelphia, Regional Science Research Institute, 105–10; and in G. Olsson (1965) *Distance and Human Interaction,* Philadelphia, Regional Science Research Institute, 8–10 and 43–68.

5 A comprehensive group of references is contained in B. H. Stevens and C. A. Brackett (1967) *Industrial Location,* Philadelphia, Regional Science Research Institute.

6 Such assumptions are integral parts of the theories of location of both Christaller and Lösch, and they are also implicit in rent theory and trade area models. For lists of relevant references see Berry and Pred, *Central Place Studies,* 97–109 (see note 4 above).

7 J. Wolpert, 'The decision process in a spatial context,' *Annals of the Association of American Geographers,* 54, 537–58.

8 P. R. Gould (1966) 'A bibliography of space-searching procedures,' Pennsylvania State University, Department of Geography, Unpublished manuscript.

9 R. G. Golledge and L. A. Brown (1967) 'Search, learning and the market decision process,' *Geografiska Annaler,* 49, 2, 116–24.

10 D. Huff (1962) 'A topological model of consumer space preferences,' *Papers of the Regional Science Association,* 6, 157–73.

11 D. Marble (1967) 'A theoretical exploration of individual travel behavior,' in W. L. Garrison and D. F. Marble (eds) *Quantitative Geography,* Evanston, Ill., Northwestern University, Studies in Geography 13.

12 A. Pred (1967) *Behavior and Location,* Lund, Gleerup.

13 G. Katona (1951) *The Psychological Analysis of Economic Behavior,* New York, McGraw-Hill, 139.

14 C. L. Hull (1964) *A Behavior System,* New York, Wiley, 228.

15 For a summary of these references see L. Brown (1968) *Diffusion Dynamics: A Review and Revision of the Quantitative Theory of the Spatial Diffusion of Innovation*, Lund, Gleerup.

16 K. Lewin (1951) *Principles of Topological Psychology*, New York, Harper.

17 K. Lewin (1935) *A Dynamic Theory of Personality*, New York, McGraw-Hill, 88–91.

18 W. Tobler (1966) 'Numerical map generalization,' Michigan Inter-University Community of Mathematical Geographers, Discussion Paper 8; A. Geits (1963) 'The determination of the location of retail activities with the use of map transformation,' *Economic Geography*, 39, 14–22; P. R. Gould (1973) 'On mental maps,' in R. M. Downs and D. Stea (eds) *Image and Environment*, Chicago, Aldine.

19 E. C. Tolman (1932) *Purposive Behavior in Animals and Man*, New York, Appleton-Century-Crofts; E. C. Tolman, B. F. Ritchie and D. Kalish (1946), 'Studies in spatial learning II: place learning versus response learning,' *Journal of Experimental Psychology*, 36, 221–9.

20 Hilgard, *Theories of Learning* (see note 2 above).

21 C. H. Hanzik and E. C. Tolman, 'The perception of spatial relations by the rat: a type of response not easily explained by conditioning,' *Journal of Comparative Psychology*, 22, 287–318.

22 E. R. Guthrie (1952) *The Psychology of Learning*, revised edn, New York, Harper; E. R. Guthrie (1942) 'Conditioning: a theory of learning in terms of stimulus, response and association,' in *The Psychology of Learning*, National Social Studies Education 41st Yearbook, 2, 17–60.

23 E. R. Guthrie (1940) 'Association and the law of effect,' *Psychological Review*, 47, 127–48.

24 W. K. Estes *et al.* (1954) *Modern Learning Theory*, New York, Appleton-Century-Crofts; W. K. Estes *et al.* (1950) 'Towards a statistical theory of learning,' *Psychological Review*, 57, 94–107.

25 Hull, *A Behavior System* (see note 14 above).

26 ibid., 6.

27 E. L. Thorndike (1932) *The Fundamentals of Learning*, New York, Teachers College. Summaries of these laws are given in Hilgard, *Theories of Learning*, 27–30 (see note 2 above).

28 Hull, *A Behavior System*, 253 (see note 14 above).

29 R. S. Yuill (1965) 'A simulation study of barrier effects in spatial diffusion problems,' Michigan Inter-University Community of Mathematical Geographers, Discussion Paper 5.

30 An extensive bibliography of relevant articles is given in Brown, *Diffusion Dynamics* (see note 15 above).

31 See Hilgard, *Theories of Learning*, 67–9 (see note 2 above).

32 R. M. Thrall, G. H. Coombs and R. L. Davis (eds) (1954) *Decision Processes*, New York, Wiley.

33 Huff, 'A toplogical model' (see note 10 above); D. J. Thompson (1966), 'Future directions in retail area research,' *Economic Geography*, 42, 1–18; Wolpert, 'The decision process in a spatial context' (see note 7 above).

34 S. Siegel *et al.* (1964) *Choice, Utility and Strategy,* New York, McGraw-Hill.

35 Throughout the sections devoted to the presentation of learning models, continued recourse was made to the useful summaries contained in R. C. Atkinson, G. H. Bowers and E. J. Crothers (1956) *An Introduction to Mathematical Learning Theory,* New York, Wiley.

36 Hull, *A Behavior System* (see note 14 above); K. W. Spence (ed.) (1960) *Behavior Theory and Learning,* Englewood Cliffs, N.J., Prentice-Hall; C. J. Burke and W. K. Estes (1957) 'A component model for stimulus variables in discrimination learning,' *Psychometrika,* 22, 133–45.

37 F. Restle (1961) *Psychology of Judgement and Choice,* New York, Wiley; M. Levine (1963) 'Mediating processes in humans at the outset of discrimination learning,' *Psychological Review,* 70, 254–76.

38 S. H. Britt (1960) *The Spenders,* New York, McGraw-Hill; L. C. Clark (ed.) (1966) *Consumer Behavior: The Dynamics of Consumer Reactions,* New York University Press; R. Ferber and H. G. Wales (1958) *Motivation and Market Behavior,* Illinois, R. D. Irwin; L. A. Fourt and J. W. Woodlock (1960) 'Early prediction of market success for new grocery products,' *Journal of Marketing,* 26; R. Frank (1962) 'Brand choice as a probability process,' *Journal of Business,* 35; A. A. Kuehn (1962) 'Consumer brand choice as a learning process,' *Journal of Advertising Research,* 2, 10–17.

39 Berry, Barnum and Tennant, 'Retail location' (see note 3 above).

40 Rushton, Golledge and Clark, 'Formulation and testing of a normative model for the spatial allocation of grocery expenditures by a dispersed population' (see note 3 above).

41 See also R. Briggs (1969) 'The scaling of preferences for spatial locations: an example using shopping centers,' Ohio State University, Department of Geography, MA thesis.

42 Expositions of this view can be found in L. Postman (1963) 'One trial learning,' in C. N. Cofer and B. S. Musgrave (eds) *Verbal Behavior and Learning,* New York, McGraw-Hill, 295–333; B. J. Underwood and G. Keppel (1962) 'One trial learning?' *Journal of Verbal Learning and Verbal Behavior,* 1, 1–13.

43 This model is expounded at length in R. R. Bush and F. Mosteller (1955) *Stochastic Models for Learning,* New York, Wiley.

44 For example, see M. F. Norman (1964) 'Incremental learning on random trials,' *Journal of Mathemetical Psychology,* 1, 336–50.

45 Bush and Mosteller, *Stochastic Models for Learning* (see note 43 above).

46 R. G. Golledge (1967), 'Conceptualizing the market decision process,' *Journal of Regional Science* 7 (supplement), 239–58.

47 For references see Stevens and Brackett, *Industrial Location* (see note 5 above).

48 M. Shubik (1959) *Strategy and Market Structure,* New York, Wiley; M. Shubik (1964) *Game Theory and Related Approaches to Social Behavior,* New York, Wiley; H. O. Nourse (1968) *Regional Economics,* New York, McGraw-Hill.

49 R. D. Luce (1958) 'A probabilistic theory of utility,' *Econometrica*, 26, 193–224; R. D. Luce, R. R. Bush and E. Galanter (eds) (1963) *Readings in Mathematical Psychology, I*, New York, Wiley; S. Siegel (1959) 'Theoretical models of choice and strategy behavior,' *Psychometrika*, 24, 306–16; H. A. Simon (1956) 'A comparison of game theory and learning theory,' *Psychometrika*, 21, 267–72.

50 See Huff, 'A topological model' (see note 10 above); Wolpert, 'The decision process in a spatial context' (see note 7 above).

51 R. D. Luce (1959) *Individual Choice Behavior: A Theoretical Analysis*, New York, Wiley; W. S. Torgerson (1958) *Theory and Methods of Scaling*, New York, Wiley; C. H. Coombs (1964) *A Theory of Data*, New York, Wiley.

52 Hull, *A Behavior System* (see note 14 above); Spence, *Behavior Theory* (see note 36 above); Siegel, 'Theoretical models' (see note 49 above); Simon, 'A comparison of game theory and learning theory (see note 49 above); P. Suppes (1961) 'Behavioristic foundations of utility,' *Econometrica*, 29, 186–202; J. B. Kruskal (1964) 'Nonmetric scaling: a numeric method,' *Psychometrika*, 29, 115–29; R. N. Shepard (1966) 'Metric structures in ordinal data,' *Journal of Mathematical Psychology*, 3, 287–315.

53 Luce, *Individual Choice Behavior* (see note 51 above).

54 This conclusion is based only on the experiments conducted at Ohio State University. Further experimentation may alter this conclusion.

55 Luce, *Individual Choice Behavior* (see note 51 above).

56 Golledge and Brown, 'Search, learning and the market decision process' (see note 9 above); G. H. Haines (1964) 'A theory of market behavior after innovation,' *Management Science*, 10, 634–55; P. Burnett (1973) 'The dimensions of alternatives in spatial choice processes,' *Geographical Analysis*, 5, 3, 181–205.

57 P. Gould (1963) 'A bibliography of space-searching procedures for geographers,' Department of Geography, Pennsylvania State University.

58 R. G. Golledge and G. Rushton (1971) 'Multidimensional scaling: review and geographic applications,' Association of American Geographers, Technical Paper 10; R. G. Golledge (1978) 'Multidimensional analysis of environmental behavior and design,' in I. Altman and J. Wohlwill (eds) *Human Behavior and Environment*, New York, Plenum Press.

59 Wolpert, 'The decision process in a spatial context (see note 7 above); P. R. Gould (1967) 'Wheat on Kilimanjaro: the perception of choice within game and learning theory frameworks,' *General Systems Yearbook 1967*, 157–66.

60 See Golledge 'Conceptualizing the market decision process,' 239–58 (see note 46 above).

61 Brown, *Diffusion Dynamics* (see note 15 above).

4 The scaling of locational preferences

Gerard Rushton

Introduction

In studies of spatial choice, many results have had questionable relevance to other areas because they reflected some of the properties of the particular spatial system from which they had been derived; the distance-decay function is one example. However, in other disciplines where models of choice have been developed, a conscious attempt has been made to construct models of preference from which conclusions can be derived that are independent of the particular set of alternatives where choices were observed. Thus a separation is made between preferences and opportunities. Models of spatial choice should attempt to define a preference scale that orders all conceivable alternatives as do many other models of choice.

The purpose of this paper is, therefore, to describe and test a methodology for finding space-preference structures from data that describe spatial choices. The methodology used is the method of

I wish to acknowledge and to thank Robert Kern, who was responsible for writing for me the computer program REVPREF, which accomplishes all of the analyses described in this paper. A listing of this program and a description of input and output forms are available on request to the Director, Computer Institute for Social Science Research, Michigan State University, East Lansing, Michigan. I would also like to acknowledge the use of J. B. Kruskal's program for multidimensional scaling and the work of the technical section of the Computer Institute for Social Science Research in adapting it to the CDC 3600 Computer. Acknowledgement is also due to Mrs Nancy Hammond for her editorial assistance.

paired comparisons, made possible by the development of a computer algorithm for searching and ordering the spatial relationship of people to alternatives, and the method of multidimensional scaling.

Spatial choice

Spatial behavior implies a search among alternatives. The criteria for an ultimate choice are presumably relative, rather than absolute, and therefore the process of spatial choice hypothesized here is one in which a person compares each alternative with every other one and selects that which he expects will give him the greatest satisfaction. Presumably, any viable model of spatial choice should, in some sense, imitate this procedure.

If the ultimate criterion of choice in this search is anticipated satisfaction, then, in constructing a model, the goal must be to express, for any possible locational choice, the degree of expected satisfaction of each alternative as a function of the relevant characteristics of the environment and of the decision maker. In view of present knowledge, this goal is far away, but a step in that direction would be the discovery of a preference function that in an ordinal way indicated which of several alternatives would give the greatest satisfaction. The extent to which decision makers have the same preference function is, of course, not known, and so it will be necessary to design tests to determine the extent to which the spatial choice patterns of different people are consistent with a single preference function. Tests for this purpose do exist.

The interest, therefore, is in passing from an individual statement of preference to the subjective-preference function. For any one individual, it is possible to derive that function, providing a large number of independent yet consistent choices have been observed. With spatial-choice data, however, it is not normally possible to observe a large number of any individual's choices. In such instances, it is assumed that the decisions of different people are generated from similar preference functions. This assumption may seem rash although it will be shown later in this paper that the measure of dissimilarity between locational types constructed from aggregate data is also a measure of interpersonal consistency.

A few recent studies have come close to establishing subjective-preference functions for spatial choice. Gould,[1] for example, has evaluated preferences for nominal locations where students were asked to rank states according to their preferences for residing there. Peterson[2] has evaluated preferences between different attributes that locations possess where subjects made decisions on the quality of neighborhoods for residential purposes. A third type of spatial preference was hypothesized by Christaller[3] where neither the names of locations nor the attributes that locations possessed were deemed relevant to spatial choice; rather, the ordering of preference was derived solely from the distance relationships between the individual and central places.

A subjective-preference function orders (that is, ranks) an individual's preferences for a set of objects. In spatial choice, the objects that require ranking are locations with different amounts of relevant properties and different spatial relations. In spatial choice, for example among towns in which a consumer makes purchases, the relevant property is related to such questions as: Does the town offer the good in question? Does it offer other goods? Does it have several outlets for the good? These are presumably independent criteria that the consumer applies to each town; but before seeking operational definitions for each, it may be noted that each measure is related to the population size of towns. Much empirical work supports the contention that large towns are more likely to offer a particular good than small towns, that they will offer more goods and services, and that they will have more establishments. Therefore, in this study the relevant properties of towns are hypothesized to be summarized in the one variable, town population. The spatial relationship of interest is the separation between the individual and the town. Although time distance is probably a more accurate operational definition of this distance than distance in miles, in the interests of simplicity and availability, this study uses distance in miles.

Thus, it is possible to take any spatial alternative facing an individual out of its unique context and to assign it to its corresponding locational type. Here locational type is defined according to its spatial relationships and the quantity of the property town size it possesses. Choices from among unique spatial alternatives are thereby considered equivalent to choices between locational types. A degree of generality has been introduced that allows a meaningful comparison to be made of decisions made in unique environmental contexts.

If locational choice can be regarded as a choice from among alternatives, then observations of spatial choice can be regarded as paired comparison data where the locational type to which the chosen location belongs is preferred to all locational types to which the rejected alternatives belong. For this conceptualization of the process of spatial choice, the only assumption made is that choice among a set of alternatives is equivalent to choice between all the paired combinations.

The method of paired comparisons

In understanding spatial choice, it is surely this unique ranking of spatial situations that is sought. Whether spatial choice is made by comparing each possibility against such a mental ranking of all conceivable opportunities and choosing that which has highest rank, or by comparing each actual alternative against every other and choosing that which gives the highest expected utility, the outcome should be the same. To imitate either process, the ranking of possible spatial situations must first be found, and this ranking can be determined from paired comparison data by techniques developed in psychology.[4] Much of the literature concerning the method of paired comparisons deals with the simple comparison of separate things, but Guttman[5] has shown that the principles involved can be extended to include the comparison of combinations of things.

> The problem of paired comparisons arises when it is desired to obtain numerical values for a set of *n* things, with respect to one characteristic, such that these values will represent the judgements of *n* individuals . . . . The judgements vary from person to person (and possibly within a person), and the problem is to determine a set of numerical values for the things being compared that will in some sense best represent or average the judgements of the whole population.
>
> In some situations, the things being compared may be simple items or objects; this we shall call the case of *ordinary* comparisons. In other situations the things may be *combinations* of items or objects.[6]

The particular least-squares estimating procedure described by

Guttman is inappropriate in practice for problems of spatial choice, since it requires that all individuals make judgements on all of the $n(n - 1)/2$ comparisons, but the method has been developed for situations where judgements on all comparisons are incomplete.[7]

Study objectives

The first objective of this paper is to derive the subjective ranking of locational types from empirical data of actual spatial choices made for a particular purpose. The second is to try to progress beyond the purely ordinal scale, which a ranking of possibilities gives us, to a cardinal measure of the subjective distance between spatial situations. A third objective is to test the resulting preference surface for consistency.

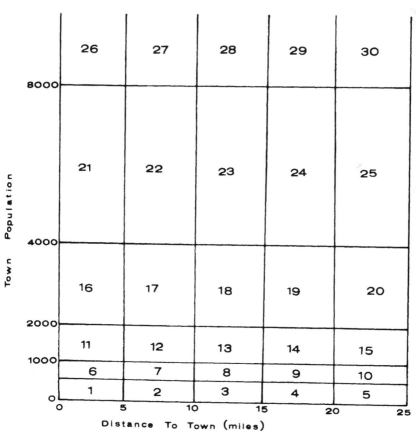

Figure 4.1   Definition of locational types

The basic matrix

The data used describe the towns selected by a random sample of rural people in Iowa for their major grocery purchases.[8] A second basic source of data is the location and size of all Iowa towns with a 1960 population greater than fifty. This forms the description of alternative opportunities among which choices were made by members of the sample.

The effect of assigning any particular town to its corresponding locational type is to take spatial choice and spatial alternatives out of their unique context. All towns within 28 miles of each member of the sample were assigned to one of thirty locational types. These types are defined in Figure 4.1. The locational type of the town chosen for major grocery purchases by the household is then considered preferred to all other locational types available. Table 4.1 shows a portion of this basic matrix. All further computations described in this work are from the data in this table.

Perceived similarity of locational types

A measure of the similarity between any two locational types is the degree to which one is preferred by persons who can choose both types. This degree can be computed from Table 4.1 as the proportion of times one type is chosen over the other when both are present. Table 4.2 shows the number of times the column locational type was preferred to the row type. The probability that the jth locational type is preferred to the ith locational type P_{jpi} can be computed from Table 4.2.[9]

Let T_{ij} be the cell total in ith row and jth column of Table 4.2 and T_{ji} be the cell total in jth row and ith column.

$$P_{jpi} = \frac{T_{ij}}{T_{ij} + T_{ji}}$$

where

$T_{ij} = 0$, and $T_{ji} = 0$, P_{jpi} and P_{ipj} are undefined

$T_{ji} = 0$, and $T_{ij} > 0$, $P_{jpi} = 1$

$T_{ij} = 0$, and $T_{ji} > 0$, $P_{jpi} = 0$

Thus, whenever

$$T_{ji} + T_{ij} > 0, P_{jpi} + P_{ipj} = 1$$

and

$$P_{jpi} = 1 - P_{ipj}$$

These probabilities are found in Table 4.3. In this table the order of both the columns and the rows represents an ordering of locational types from the preference data according to the ratio in the last column of this table. Each number in this column shows the proportion of times the locational type for the row was preferred to the other types.

Maximum perceived similarity between any two locational types, i and j, would have the value 0.5 in Table 4.3; that is: $P_{jpi} = 0.5$ and $P_{ipj} = 0.5$. Any departure from this value in either direction represents an increase in perceived dissimilarity between the locational types in question. A measure of perceived similarity d_{ij} (distance in perceived dissimilarity between locational types i and j), is therefore given by:

$$d_{ij} = |P_{jpi} - 0.5|$$

Scaling the similarity measures

From these proximity measures a scale is to be constructed on which all locational types will be positioned. Interpoint distances measured from this scale ought to correspond to the original proximity measures. Such a scale would, of course, order the locational types and thus would demonstrate that the spatial-choice data can be regarded as having been generated from a preference structure, since preference structures are an ordering of all conceivable alternatives. The particular preference structure that generated the data set would thus have been identified. Knowing that from it a large proportion of spatial choices can be predicted would confirm the conceptualization of the spatial-choice problem outlined earlier. From the scale, measures can be made of the dissimilarity between locational types for which no data were present in the original matrix of interpoint dissimilarities.[10] This possibility has far-reaching implications for the analysis of choices

Table 4.1 *Revealed space-preference–raw-data matrix by respondents*

Household							Locational type																							
ID	1	2	3	4	5	6	7	8	9	10	11	12	13	14	15	16	17	18	19	20	21	22	23	24	25	26	27	28	29	30
1	*	*	*	*	*				*	*			*		*		1								*					
2	*	*	*	*	*	*			*	*	*			*					1											
3	*	*	*	*	*	*		*	*	*	*		1	*					*											
3	*	*	*	*	*	*		*	*	*		1	1	*	*			*												
4	*	*	*	*	*	*		*	*	*	1				*	1														
6	*	*	*	*	*	*	*	*	*	*			*	*	*			*	*	*			1		*					
9	*	*	*	*			*		*										*	*			1							
10	*	*	*	*					*	*						*			*	*			1							
11	*	*	*				*	*	*	*								*	*	*			1							
12	*	*	1	*			*	*	*									*	*	*	*	1								
13	*	*	*	*		*	*	*	*										*	*										
14	*	*	*	*	*	1	*	*	*										*											
15	*	*	*	*	1	1	*	*	*	*							*													
16	*	*	*			*	*	*	*																					
18	*	*	*	*		*	*	*								1														

1 Locational type patronized

* Locational type rejected

Blank Locational type not present

Source: Computed from data described in note 8 in this chapter (only part of table shown)

Table 4.2 *Segment of the revealed-preference data matrix* (cells show number of times sample households preferred column locational type to row type)

Household	Locational type																	
	1	2	3	4	5	6	7	8	9	10	11	12	13	14	15	16	17	18
1	29.8	9.0	3.0	1.0	0.0	27.0	15.3	3.0	1.0	0.0	23.5	47.0	7.2	0.0	0.0	22.0	33.5	25.9
2	127.7	21.5	2.0	2.0	0.0	90.5	55.3	12.0	0.0	0.0	85.0	94.3	23.5	0.0	0.0	81.0	77.5	65.7
3	180.0	27.5	11.0	0.0	0.0	188.5	97.5	22.0	3.0	0.0	176.5	154.2	25.0	0.0	0.0	119.0	133.0	101.5
4	243.3	36.0	8.0	9.0	0.0	228.0	137.3	14.0	8.0	0.0	211.5	190.5	44.0	0.0	0.0	126.0	193.0	106.8
5	209.8	37.0	8.0	5.0	0.0	223.0	133.3	18.0	8.0	0.0	224.0	193.7	47.5	0.0	0.0	165.0	180.5	100.7
6	5.5	0.0	1.0	0.0	0.0	4.0	5.0	0.0	0.0	0.0	4.0	6.5	1.0	0.0	0.0	2.0	7.5	4.0
7	30.5	5.5	1.0	0.0	0.0	36.5	17.3	1.0	2.0	0.0	14.5	24.0	5.0	0.0	0.0	12.0	20.5	10.0
8	41.7	7.0	1.0	1.0	0.0	41.0	33.0	3.0	1.0	0.0	40.0	31.8	6.2	0.0	0.0	27.0	32.5	19.2
9	66.7	5.0	4.0	0.0	0.0	63.0	34.0	2.0	0.0	0.0	51.0	54.3	12.7	0.0	0.0	45.0	54.5	15.7
10	65.8	10.5	2.0	2.0	0.0	51.5	34.8	5.0	0.0	0.0	65.5	56.8	10.0	0.0	0.0	36.5	53.0	27.8
11	2.5	0.0	0.0	0.0	0.0	1.0	0.0	0.0	0.0	0.0	1.0	0.0	1.5	0.0	0.0	0.0	4.0	1.5
12	8.5	3.0	0.0	0.0	0.0	10.0	3.0	0.0	0.0	0.0	5.0	8.3	1.0	0.0	0.0	8.0	10.0	4.0
13	23.3	1.5	0.0	1.0	0.0	24.5	17.0	0.0	1.0	0.0	24.5	18.5	7.5	0.0	0.0	8.0	23.0	10.3

14	32.8	6.0	0.0	1.0	0.0	0.0	46.5	16.5	2.0	1.0	0.0	0.0	29.5	28.3	3.8	0.0	0.0	17.5	17.0	9.8
15	43.8	3.0	0.0	1.0	0.0	0.0	35.0	15.3	0.0	0.0	0.0	0.0	27.0	37.3	9.5	0.0	0.0	23.5	22.0	11.3
16	0.0	0.0	0.0	0.0	0.0	0.0	0.0	0.0	0.0	0.0	0.0	0.0	1.0	0.0	0.0	0.0	0.0	2.0	0.0	1.0
17	3.0	0.5	0.0	0.0	0.0	0.0	8.0	4.5	1.0	0.0	0.0	0.0	6.0	6.0	1.0	0.0	0.0	1.0	6.5	1.0
18	16.6	3.0	2.0	1.0	0.0	0.0	13.5	6.3	1.0	2.0	0.0	0.0	7.5	3.0	2.3	0.0	0.0	3.0	3.5	4.5
19	16.5	0.5	3.0	0.0	0.0	0.0	25.0	10.0	2.0	0.0	0.0	0.0	20.5	19.0	3.0	0.0	0.0	11.0	17.5	3.5
20	9.3	4.5	1.0	0.0	0.0	0.0	12.0	10.3	2.0	0.0	0.0	0.0	17.5	20.0	2.5	0.0	0.0	6.0	16.0	5.8
21	0.0	0.0	0.0	0.0	0.0	0.0	2.0	0.0	0.0	0.0	0.0	0.0	0.0	0.0	1.0	0.0	0.0	0.0	0.0	0.0
22	3.0	0.0	0.0	0.0	0.0	0.0	6.0	2.0	0.0	0.0	0.0	0.0	1.0	2.0	1.0	0.0	0.0	1.0	1.0	0.0
23	13.0	1.5	1.0	0.0	0.0	0.0	5.0	5.0	0.0	0.0	0.0	0.0	8.0	5.0	2.0	0.0	0.0	5.0	1.5	1.0
24	16.0	2.0	1.0	0.0	0.0	0.0	8.5	6.5	1.0	0.0	0.0	0.0	13.5	8.0	2.0	0.0	0.0	9.0	7.5	2.0
25	14.3	2.0	0.0	0.0	0.0	0.0	11.0	8.3	3.0	0.0	0.0	0.0	16.5	12.0	2.0	0.0	0.0	10.0	7.5	3.8
26	0.0	0.0	0.0	0.0	0.0	0.0	0.0	0.0	0.0	0.0	0.0	0.0	0.0	0.0	0.0	0.0	0.0	2.0	1.0	1.5
27	2.0	0.0	0.0	1.0	0.0	0.0	1.0	1.0	0.0	1.0	0.0	0.0	2.0	1.0	0.0	0.0	0.0	0.0	0.0	0.0
28	2.0	1.0	0.0	0.0	0.0	0.0	5.5	3.0	0.0	0.0	0.0	0.0	4.0	2.0	1.0	0.0	0.0	1.0	1.0	0.5
29	7.3	0.0	0.0	0.0	0.0	0.0	13.0	7.3	0.0	0.0	0.0	0.0	8.5	5.0	0.0	0.0	0.0	6.0	6.0	2.8
30	13.3	1.0	0.0	0.0	0.0	0.0	4.5	5.3	0.0	0.0	0.0	0.0	9.0	9.3	2.0	0.0	0.0	9.5	4.0	3.8

Source: Computed from data described in note 8 in this chapter (only part of table shown)

Table 4.3   Probability that column locational type is preferred to row type

	21	16	27	11	26	28	22	23	17	29	6	12	18	24	1	7
21	-1.00	-1.00	-1.00	-1.00	-1.00	-1.00	-1.00	0.00	-1.00	-1.00	0.25	0.00	0.00	0.00	0.00	0.00
16	-1.00	-1.00	-1.00	-1.00	0.00	0.00	0.00	0.17	0.00	0.14	0.00	0.00	0.25	0.00	0.00	0.00
27	-1.00	-1.00	-1.00	0.40	1.00	0.25	0.00	0.00	0.00	0.00	0.11	0.20	0.00	0.00	0.06	0.05
11	-1.00	0.00	0.60	-1.00	-1.00	0.56	0.00	0.11	0.40	0.35	0.20	0.00	0.17	0.04	0.10	0.00
26	-1.00	1.00	0.00	-1.00	-1.00	0.00	0.00	0.00	0.50	-1.00	0.00	0.00	0.50	0.00	0.00	0.00
28	-1.00	1.00	0.75	0.44	1.00	-1.00	1.00	0.67	0.13	0.00	0.42	0.22	0.10	0.00	0.05	0.10
22	-1.00	1.00	1.00	1.00	1.00	0.00	-1.00	0.00	1.00	0.25	0.35	0.17	0.00	0.06	0.07	0.09
23	1.00	0.83	1.00	0.89	1.00	0.33	1.00	-1.00	0.38	0.57	0.37	0.27	0.08	0.27	0.24	0.17
17	-1.00	1.00	1.00	0.60	0.50	0.88	0.00	0.63	-1.00	0.00	0.52	0.38	0.22	0.06	0.08	0.18
29	-1.00	0.86	1.00	0.65	-1.00	1.00	0.75	0.43	1.00	-1.00	0.81	0.45	0.29	0.00	0.30	0.38
6	0.75	1.00	0.89	0.80	1.00	0.58	0.65	0.63	0.48	0.19	-1.00	0.39	0.23	0.32	0.17	0.12
12	1.00	1.00	0.80	1.00	1.00	0.78	0.83	0.73	0.63	0.55	0.61	-1.00	0.57	0.34	0.15	0.11
18	1.00	0.75	1.00	0.83	0.50	0.90	1.00	0.92	0.78	0.71	0.77	0.43	-1.00	0.71	0.39	0.38
24	1.00	1.00	1.00	0.96	1.00	1.00	0.94	0.73	0.94	1.00	0.68	0.66	0.29	-1.00	0.50	0.39
1	1.00	1.00	0.94	0.90	1.00	0.95	0.93	0.76	0.92	0.70	0.83	0.85	0.61	0.50	-1.00	0.33
7	1.00	1.00	0.95	1.00	1.00	0.90	0.91	0.83	0.82	0.62	0.88	0.89	0.62	0.61	0.67	-1.00
13	0.92	1.00	1.00	0.94	1.00	0.93	0.97	0.88	0.96	1.00	0.96	0.95	0.82	0.67	0.77	0.77
30	1.00	1.00	1.00	1.00	1.00	0.88	1.00	0.81	0.73	0.73	1.00	0.78	0.71	0.57	0.75	0.78
25	1.00	1.00	1.00	1.00	1.00	0.90	1.00	1.00	1.00	1.00	0.92	0.92	1.00	0.67	0.83	0.73
20	1.00	1.00	1.00	1.00	1.00	1.00	1.00	1.00	1.00	0.85	0.92	1.00	0.85	1.00	0.90	0.77
8	1.00	1.00	1.00	1.00	1.00	1.00	1.00	1.00	0.97	1.00	1.00	1.00	0.95	0.86	0.93	0.97
2	1.00	1.00	1.00	1.00	1.00	0.99	1.00	0.99	0.99	1.00	1.00	0.97	0.96	0.94	0.93	0.91
19	1.00	1.00	1.00	1.00	1.00	1.00	1.00	1.00	1.00	1.00	0.96	1.00	1.00	-1.00	0.85	0.91
3	1.00	1.00	1.00	0.95	1.00	1.00	1.00	0.99	1.00	1.00	0.99	1.00	0.98	0.98	0.98	0.99
4	1.00	1.00	0.99	1.00	1.00	1.00	1.00	1.00	1.00	1.00	1.00	1.00	0.99	1.00	1.00	1.00
9	1.00	1.00	0.98	1.00	1.00	1.00	1.00	1.00	1.00	1.00	1.00	1.00	0.89	1.00	0.99	0.94
15	1.00	1.00	1.00	1.00	1.00	1.00	1.00	1.00	1.00	1.00	1.00	1.00	1.00	1.00	1.00	1.00
14	1.00	1.00	1.00	1.00	1.00	1.00	1.00	1.00	1.00	1.00	1.00	1.00	1.00	1.00	1.00	1.00
10	1.00	1.00	1.00	1.00	1.00	1.00	1.00	1.00	1.00	1.00	1.00	1.00	1.00	1.00	1.00	1.00
5	1.00	1.00	1.00	1.00	1.00	1.00	1.00	1.00	1.00	1.00	1.00	1.00	1.00	1.00	1.00	1.00

	13	30	25	20	8	2	19	3	4	9	15	14	10	5	Per cent >0.5
21	0.08	0.00	0.00	0.00	0.00	0.00	0.00	0.00	0.00	0.00	0.00	0.00	0.00	0.00	100.00
16	0.00	0.00	0.00	0.00	0.00	0.00	0.00	0.00	0.00	0.00	0.00	0.00	0.00	0.00	96.30
27	0.00	0.00	0.00	0.00	0.00	0.00	0.00	0.00	0.01	0.02	0.00	0.00	0.00	0.00	96.30
11	0.06	0.00	0.00	0.00	0.00	0.00	0.05	0.00	0.00	0.00	0.00	0.00	0.00	0.00	92.59
26	0.00	0.00	0.00	0.00	0.00	0.00	0.00	0.00	0.00	0.00	0.00	0.00	0.00	0.00	88.46
28	0.07	0.13	0.10	0.00	0.00	0.01	0.00	0.00	0.00	0.00	0.00	0.00	0.00	0.00	82.14
22	0.03	0.00	0.00	0.00	0.00	0.00	0.00	0.01	0.00	0.00	0.00	0.00	0.00	0.00	82.14
23	0.13	0.19	0.00	0.00	0.00	0.01	0.00	0.00	0.00	0.00	0.00	0.00	0.00	0.00	75.86
17	0.04	0.27	0.00	0.15	0.03	0.01	0.00	0.00	0.00	0.00	0.00	0.00	0.00	0.00	75.00
29	0.00	0.27	0.00	0.08	0.00	0.00	0.00	0.00	0.00	0.00	0.00	0.00	0.00	0.00	74.07
6	0.04	0.00	0.08	0.00	0.00	0.00	0.04	0.01	0.00	0.00	0.00	0.00	0.00	0.00	72.41
12	0.05	0.22	0.08	0.00	0.00	0.03	0.00	0.00	0.00	0.00	0.00	0.00	0.00	0.00	58.62
18	0.18	0.29	0.00	0.15	0.05	0.04	0.00	0.02	0.01	0.00	0.00	0.00	0.00	0.00	58.62
24	0.33	0.43	0.33	0.00	0.14	0.06	−1.00	0.02	0.00	0.11	0.00	0.00	0.00	0.00	53.57
1	0.23	0.25	0.17	0.10	0.07	0.07	0.15	0.02	0.00	0.00	0.00	0.00	0.00	0.00	51.72
7	0.23	0.22	0.27	0.23	0.03	0.09	0.09	0.01	0.00	0.01	0.00	0.00	0.00	0.00	48.28
13	−1.00	0.20	0.00	0.00	0.00	0.06	0.00	0.00	0.02	0.06	0.00	0.00	0.00	0.00	44.83
30	0.80	−1.00	0.67	−1.00	0.00	0.11	0.00	0.00	0.00	0.07	0.00	0.00	0.00	0.00	35.71
25	1.00	0.33	−1.00	−1.00	0.38	0.50	0.00	0.00	0.00	0.00	−1.00	0.00	0.00	0.00	33.33
20	1.00	−1.00	−1.00	−1.00	0.40	0.60	0.33	0.10	0.00	0.00	0.00	0.00	−1.00	0.00	30.77
8	1.00	1.00	0.63	0.60	−1.00	0.37	0.33	0.04	0.07	0.33	−1.00	0.00	0.00	0.00	28.57
2	0.94	0.89	0.50	0.40	0.63	−1.00	0.86	0.07	0.05	0.00	0.00	0.00	0.00	0.00	27.59
19	1.00	1.00	1.00	0.67	0.67	0.14	−1.00	0.25	0.00	0.00	0.00	−1.00	−1.00	0.00	23.08
3	1.00	1.00	1.00	0.90	0.96	0.93	0.75	−1.00	0.00	0.43	0.00	−1.00	0.00	0.00	14.81
4	0.98	1.00	1.00	1.00	0.93	0.95	1.00	1.00	−1.00	1.00	−1.00	0.00	0.00	0.00	13.79
9	0.93	1.00	1.00	1.00	0.67	1.00	1.00	0.57	0.00	−1.00	−1.00	−1.00	−1.00	−1.00	11.11
15	1.00	1.00	−1.00	1.00	−1.00	1.00	1.00	−1.00	1.00	−1.00	−1.00	−1.00	−1.00	−1.00	0.00
14	1.00	1.00	1.00	1.00	1.00	1.00	−1.00	−1.00	1.00	−1.00	−1.00	−1.00	−1.00	−1.00	0.00
10	1.00	1.00	1.00	−1.00	1.00	1.00	−1.00	1.00	1.00	−1.00	−1.00	−1.00	−1.00	−1.00	0.00
5	1.00	1.00	1.00	1.00	1.00	1.00	1.00	1.00	1.00	−1.00	−1.00	−1.00	−1.00	−1.00	0.00

Source: Computed from Table 4.2 (−1.00 is missing data)

in an area where the distribution of alternative opportunities differs substantially from that of the area where the sample data were gathered. Another reason for searching for a scale is that, in the conceptualization of the decision-making process for spatial choice, the conceptualization is of a rule being present whereby all alternative spatial opportunities can be judged and choice made. Such a rule would imply that considerably more than ordinal information on the location of points in a space exists, for otherwise the rank order of distances between all conceivable points could not be known. Hence spatial choice appears to be an example of a situation where '*ordinal* information on distances does imply a considerable amount of *interval* information on the location of the points.'[11]

The technique described below is based on a ranking of the measures of dissimilarity. Thus qualities of additivity to the measures of dissimilarity are not necessarily imputed. Since they are derived from ordinal relationships (from the paired comparisons), an intrinsic-interval level of measurement in the data matrix cannot be assumed even though it is assumed that such a scale must have been present for the matrix to be generated. However, Shepard's work has shown the possibility, in certain specified circumstances, of deriving from ordinal-type data approximate metric scales that have a high degree of accuracy.

Let δ_{ij} be any one of the measures of dissimilarity, described above.[12] For a matrix of such dissimilarities the intent is to represent the n locational types as n points in t dimensional space, the interpoint distances of which (d_{ij}) correspond to the observed degree of dissimilarity between the n locational types. Perfect correspondence would mean, for example, that, if locational type i is more similar to type j than it is to type k, then the corresponding interpoint distances would satisfy the same relationship – for all i, j, k. That is, where $\delta_{ij} > \delta_{ik} : d_{ij} > d_{ik}$. The simplest example would be one in which locational types could be so arranged in one-dimensional space that the ranking of interpoint distances corresponded to the ranking of dissimilarities in the probability matrix.

With n locational types, assuming complete data, there are $n(n-1)/2$ dissimilarities. Ignoring, for the moment, the possibilities of ties, the dissimilarities can be ranked in ascending order.

$$\delta_{i_1 j_1} < \delta_{i_2 j_2} < \delta_{i_3 j_3} < \ldots \delta_{i_M j_M} \ldots$$

Here $M = n(n-1)/2$. Let the locational types be called $x_i, \ldots, x_n$

and expressed in orthogonal coordinates by $x_i = (x_{i1}, \ldots, x_{i3}, \ldots, x_{it})$. Let d_{jk} denote the distance from x_j to x_k; then

$$d_{jk} = \left[\sum_{r=1}^{n} (P_{rj} - P_{rk})^2 \right]$$

where

j, k are indices for any two points
r is an index for axes
n is the number of orthogonal axes
P_{rj} refers to the projection of point j on axis r

Correspondence between the interpoint distances and the dissimilarities would mean that if the distances were ranked from smallest to largest then the same order of the locational types is maintained.

That is:

$$d_{i_1j_1} < d_{i_2j_2} < d_{i_3j_3} < \ldots \ldots \ldots d_{iMjM}$$

In other words, if the locational types are shown on a scatter plot in which the ordinate is dissimilarity (δ) and the abscissa is distance (d), then as the points are traced one by one from bottom to top, the move is always to the right, never to the left. When this requirement has been met, a monotone relationship between dissimilarity and distance has been found.

With empirical data, however, perfect correspondence is rarely achieved and so some measure of goodness of fit is required. Kruskal proposes a test similar to the correlation coefficient that he calls stress:

$$S = \sqrt{\left(\frac{\sum_{i<j} (d_{ij} - \hat{d}_{ij})^2}{\sum_{i<j} d_{ij}^2} \right)}$$

Where $\hat{d}_{ij}$ is the minimum distance between x_i and x_j, that will satisfy the monotonic relationship.

Kruskal has developed an algorithm for finding the orthogonal coordinates for the n points such that for any number of dimensions stress is minimized. When some of the dissimilarities are missing the terms that correspond to the missing dissimilarities are dropped in both the numerator and the denominator of the definition of stress.

Table 4.4   *Scale values for the locational types*

Rank	Locational type	Scale value
1	21	−1.311
2	16	−1.523
3	27	−1.382
4	11	−0.991
5	26	−1.624
6	28	−0.876
7	22	−1.160
8	23	−0.590
9	17	−0.719
10	29	−0.668
11	6	−0.725
12	12	−0.481
13	18	−0.321
14	24	−0.020
15	1	−0.163
16	7	−0.075
17	13	−0.217
18	30	−0.009
19	25	0.378
20	20	0.532
21	8	0.752
22	2	0.551
23	19	0.812
24	3	1.114
25	4	1.174
26	9	1.266
27	15	1.438
28	14	1.596
29	10	1.517
30	5	1.725

Stress = 0.434

Source: Computed according to algorithm described in J. B. Kruskal (1964) 'Nonmetric multidimensional scaling: a numerical method,' *Psychometrika*, 29, 115–29.

The extent to which spatial choice can be understood as the application of a subjective ranking of locational types to the particular unique set of spatial alternatives facing an individual is described by the amount of stress on the first dimension. Table 4.4 shows the computed scale values on the first dimension. In Figure 4.2, the scales are shown as isopleths of this data. These interpolated lines are of equal scale value on the two variables, population size and distance, that were used in defining the locational types. This surface is called an indifference surface of spatial choice, because it has all of the typical features of a preference surface. The inference to be made is that a person would be indifferent between (that is, expect equal satisfaction from) any two towns placed along one of the isolines, and would prefer (that is, expect most satisfaction) from the town that lies on the highest point of the surface.

One feature of the scale developed is that it is frequently not consistent with the replies of respondents with respect to individual pairs of locational types. For example, although type 16 is shown to give more satisfaction than type 11 on the final scale (Figure 4.2),

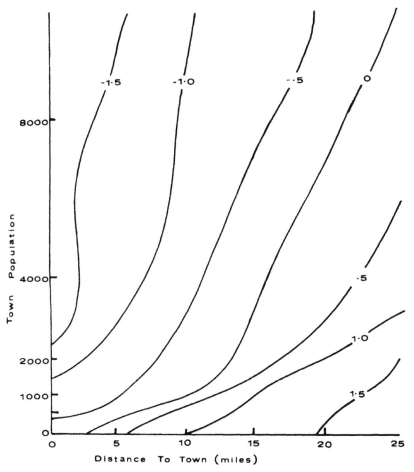

Figure 4.2 Space preference structure for grocery purchases; Iowa, 1960

Table 4.3 shows that on the occasions when a person made a choice between types 11 and 16, type 11 was always preferred. However, Table 4.2 shows that there was only one such occasion and it is therefore not surprising that in such circumstances sampling error is large. Indeed, opportunities for comparison between locational types that are close on the satisfaction scale might be uncommon, and, accordingly, large sampling variations will surround any values that do occur. One advantage of the scaling method employed here is that it uses information in addition to the comparison of any two locational types to determine the distance between them on the final scale. In fact, the analysis has shown that, given sufficient qualitative statements of the form 'locational type i is preferred to locational type j,' it is possible to derive a metric scale that gives a meaningful distance measure between any two locational types. The conclusion might seem to contradict a time-honored distinction between ordinal and interval measurement scales; but recent theoretical and experimental results using both the Kruskal scaling technique used in this paper and other techniques have shown that, given a sufficient number of inequalities on the interpoint distances, the location of points in any given number of dimensions is free to move only within narrow limits before some interpoint inequality is broken.[13] Thus, in certain circumstances, metric scales with a very high degree of accuracy can be developed from purely ordinal data.

Transitivity and the scale

A more conventional test of whether this assumption of unidimensionality in the judgement between locational types is justified, is the extent to which the matrix of probabilities is fully transitive. This test is commonly referred to in the psychological literature as the test for weak stochastic transitivity. In Table 4.5, the data from Table 4.3 is transformed into a binary matrix by substituting a 1 whenever $P_{jpi} > 0.5$ and a 0 whenever $P_{jpi} < 0.5$.

The coefficient of consistency for such a matrix varies from 0 (maximum inconsistency) to 1 (complete transitivity). The 1 would indicate that a unidimensional preference structure obtains.[14] The coefficient can be described verbally as one minus the ratio between the observed number of cyclic triplets in the matrix to the maximum number of possible cyclic triplets in the matrix, where a cyclic triplet is an intransitivity. An intransitivity can be described as the example

Table 4.5   *Transitivity test for consistency of preference surface*

Locational types

	21	16	27	11	26	28	22	23	17	29	6	12	18	24	1	7	13	30	25	20	8	2	19	3	4	9	15	14	10	5
21	0	2	2	2	2	2	2	0	2	2	0	0	0	0	0	0	0	0	0	0	0	0	0	0	0	0	0	0	0	0
16	2	0	2	1	0	0	0	0	0	0	0	0	0	0	0	0	0	0	0	0	0	0	0	0	0	0	0	0	0	0
27	2	2	0	0	1	0	0	0	0	0	0	0	0	0	0	0	0	0	0	0	0	0	0	0	0	0	0	0	0	0
11	2	0	1	0	2	1	0	0	0	0	0	0	0	0	0	0	0	0	0	0	0	0	0	0	0	0	0	0	0	0
26	2	1	0	2	0	0	0	0	2	2	0	0	2	0	0	0	0	0	0	0	0	0	0	0	0	0	0	0	0	0
28	2	1	1	0	1	0	1	1	0	0	0	0	0	0	0	0	0	0	0	0	0	0	0	0	0	0	0	0	0	0
22	2	1	1	1	1	0	0	0	1	0	0	0	0	0	0	0	0	0	0	0	0	0	0	0	0	0	0	0	0	0
23	1	1	1	1	1	0	1	0	0	1	0	0	0	0	0	0	0	0	0	0	0	0	0	0	0	0	0	0	0	0
17	2	1	1	1	2	1	0	1	0	0	1	0	0	0	0	0	0	0	0	0	0	0	0	0	0	0	0	0	0	0
29	2	1	1	1	2	1	1	0	1	0	1	0	0	0	0	0	0	0	0	0	0	0	0	0	0	0	0	0	0	0
6	1	1	1	1	1	1	1	1	1	0	0	0	0	0	0	0	0	0	0	0	0	0	0	0	0	0	0	0	0	0
12	1	1	1	1	1	1	1	1	1	1	1	0	1	0	0	0	0	0	0	0	0	0	0	0	0	0	0	0	0	0
18	1	1	1	1	2	1	1	1	1	1	1	0	0	1	0	0	0	0	0	0	0	0	0	0	0	0	0	0	0	0
24	1	1	1	1	1	1	1	1	1	1	1	1	0	0	2	0	0	0	0	0	0	0	2	0	0	0	0	0	0	0
1	1	1	1	1	1	1	1	1	1	1	1	1	1	1	2	0	0	0	0	0	0	0	0	0	0	0	0	0	0	0
7	1	1	1	1	1	1	1	1	1	1	1	1	1	1	1	0	0	0	0	0	0	0	0	0	0	0	0	0	0	0
13	1	1	1	1	1	1	1	1	1	1	1	1	1	1	1	1	0	0	0	0	0	0	0	0	0	0	0	0	0	0
30	1	1	1	1	1	1	1	1	1	1	1	1	1	1	1	1	1	0	1	2	0	0	0	0	0	0	0	0	0	0
25	1	1	1	1	1	1	1	1	1	1	1	1	1	1	1	1	1	1	0	0	2	0	2	0	0	0	2	0	0	0
20	1	1	1	1	1	1	1	1	1	1	1	1	1	1	1	1	1	2	2	0	0	1	0	0	0	0	0	0	2	0
8	1	1	1	1	1	1	1	1	1	1	1	1	1	1	1	1	1	1	1	1	0	0	0	0	0	0	2	0	0	0
2	1	1	1	1	1	1	1	1	1	1	1	1	1	1	1	1	1	1	1	1	2	0	1	0	1	0	0	0	0	0
19	1	1	1	1	1	1	1	1	1	1	1	1	1	2	1	1	1	1	1	1	1	1	0	0	0	0	0	2	2	0
3	1	1	1	1	1	1	1	1	1	1	1	1	1	1	1	1	1	1	1	1	1	1	1	0	0	0	2	2	0	0
4	1	1	1	1	1	1	1	1	1	1	1	1	1	1	1	1	1	1	1	1	1	1	1	1	0	1	0	0	0	0
9	1	1	1	1	1	1	1	1	1	1	1	1	1	1	1	1	1	1	1	1	1	1	1	1	0	0	2	0	2	0
15	1	1	1	1	1	1	1	1	1	1	1	1	1	1	1	1	1	1	2	1	2	1	1	2	1	2	0	2	2	2
14	1	1	1	1	1	1	1	1	1	1	1	1	1	1	1	1	1	1	1	1	1	1	2	2	1	1	2	0	2	2
10	1	1	1	1	1	1	1	1	1	1	1	1	1	1	1	1	1	1	2	1	1	2	1	1	1	2	2	2	0	2
5	1	1	1	1	1	1	1	1	1	1	1	1	1	1	1	1	1	1	1	1	1	1	1	1	1	1	2	2	2	0

1 Column locational type is preferred to row type
2 Missing data
Source: Computed from Table 4.3

in which locational type i is preferred to j, type j is preferred to k, and yet type k is revealed as preferred to i. Clearly, the greater the proportion of such cyclic triplets present of all such triplets that could be present, the less one can speak of a unidimensional general preference structure. The lower the coefficient of consistency the more likely that several distinct preference structures exist in the sample population studied. The coefficient of consistency[15] was computed from Table 4.5. It has a value of 0.985.

Some unsolved problems

The problem of aggregation

To the extent that the proximity measures are measures of how frequently the population of individuals disagreed about a common scale, they provide evidence on the important subject of inter-personal consistency of spatial behavior. Where, for example, location type A is always preferred to type B, the scale that represents A as higher than B is common to all individuals for that comparison. Where A was preferred to B as frequently as B was preferred to A, the two points occupy the same position on the common scale although on the individuals' scales, A and B may well be separated. Coombs points the way to future research here when he writes:

> An experimentor may be interested, not in constructing a common stimulus scale for a population of individuals, but rather in partitioning a population into subgroups, each of which has a common but distinctive stimulus scale. Then the intent is to find the distinguishing characteristics of the subpopulations which are associated with the corresponding distinctions between stimulus scales.[16]

The problem of temporal changes

In addition to the question of how similar the preference structure of one person is to that of another is the question of how stable a person's preference structure is through time. Since the individual operates in a constantly changing spatial environment, one aspect of spatial preference is presumably the rules by which a person acquires knowledge about the environment and, indeed, antici-pates changes in it. Curry and others[17] have recently emphasized the interdependence, and consequently the mutual adaptation, of the activities of consumers and retailers in a continuous learning process. Models are needed 'to separate out the momentary fluctuations in behavior which may obscure the behavior's more persistent and organized character.'[18]

The problem of aggregation and of temporal change can be shown as the combination of two response matrices. The first one (Table 4.6) illustrates the possibility of different individuals, responding differently to the same set of stimuli (locational types) on a single occasion, while the second matrix (Table 4.7) refers to responses by the same individual on different occasions. Combining the two, a box can be conceived of with responses related to the three axes of locational types, individuals and occasions. The

Table 4.6   *A response matrix showing responses to various locational types by various individuals on a single occasion*

Individuals	Locational types				
	L_1	L_2	L_3	L_j	L_n
I_1	R_{11}	R_{12}		R_{ij}	R_{1n}
I_2	R_{21}	R_{22}		R_{2j}	R_{2n}
I_3	R_{31}	R_{32}		R_{3j}	R_{3n}
,					
,					
,					
I_i	R_{i1}	R_{i2}		R_{ij}	R_{in}
,					
,					
I_N	R_{N1}	R_{N2}		R_{Nj}	R_{Nn}

spatial-learning studies referred to above become, in this context, a study of individuals' patterns of responses among occasions with a view to determining whether revealed inconsistencies in a unitary preference structure by the same individual are the momentary fluctuations that Coombs refers to, or whether they represent a systematic change in the preference structure itself. That spatial-choice patterns change through time is neither sufficient nor necessary to prove that preference structures have changed. Changes in spatial choice patterns result from changes in the distribution of alternatives, changes in the 'action space' of individuals through the learning process, as well as changes in the individual's preference structure. These three influences must be identified and separated for a meaningful study of spatial behavior.

Table 4.7   *A response matrix showing responses to various locational types on various occasions by the same individual*

Occasions	Locational types				
	L_1	L_2	L_3	L_j	L_n
0_1	R_{11}	R_{12}	R_{13}	R_{1j}	R_{1n}
0_2	R_{21}			R_{2j}	R_{2n}
0_3	R_{31}			R_{3j}	R_{3n}
,					
,					
,					
0_i	R_{i1}			R_{ij}	R_{in}
,					
,					
0_n	R_{N1}			R_{Nj}	R_{Nn}

Presumably, 'action space' is related to the distribution of alternatives, whereas preference structures are independent of any particular distribution of alternatives. How else could spatial choices be made in new environments? However, the question of how preference structures change through time, and how stable they are through time still remains.

The problem of surrogates

The operational definition of locational types used in this study was influenced by the availability of data, but the conceptualization of the scaling problem described here is independent of these definitions. The problem of surrogates is the problem of using intermediate entities as operational definitions of the more fundamental properties that presumably one is interested in.[19] Houthaker discusses this problem in relation to indifference maps in economics in which, he argues, even the quantities described are presumably surrogates for more abstract properties in several dimensions that each quantity possesses.[20] It is not uncommon for it to be impossible to derive independent measures of the separate factors that influence behavior; only the order of their joint effect is known. The problem then becomes one of finding the measurement scales both for the factors and for their effects. This problem is known as the conjoint measurement problem and has recently received widespread discussion.[21]

The problem of error

The scaling model described in this paper is a deterministic model. No provision is made for any error or aberrant response, and, since such responses may occur, deviation between actual and model behavior is to be expected. Unlike probabilistic type-scaling models, such models present problems in evaluating goodness of fit.[22] Furthermore, sampling error will be important when the number of locational types is large relative to the number of revealed preferences. Guilford suggests that 'possibly a good rule would be to limit application to where N/n is greater than fifty.'[23]

One writer has recently argued with impressive supporting evidence that, whenever choice has to be made among alternatives, each of which consists of a number of subjectively disparate attributes, man's ability 'to arrive at an overall evaluation by weighting and combining or "trading off" all of these separate attributes'[24] is not impressive; that is, it can frequently be shown to be inconsistent with his stated preferences in evaluating the alternatives on any one of the subjective attributes. Shepard describes the results of experiments that indicate that man often fails in his efforts to combine correctly the effects of the multitude of factors that influence his behavior and so fails to choose from the set of alternatives the one that would give him the greatest satisfaction.

Summary

The study of spatial choice should be a study of revealed preference between locational types. As paired comparisons between locational types, such data can be used to establish a preference function from which it is possible to derive the subjective ranking of locational types. By using recent advances in scaling techniques, it is possible to derive a measure of the perceived dissimilarity between locational types. Finally, a test for stochastic transitivity in the revealed-preference matrix reveals the extent to which the assumption of a unidimensional preference structure is valid.

The purpose of studying spatial behavior should be to derive a description of preferences. Opportunities for spatial interaction are readily observable, and consequently spatial choice is predictable by placing these opportunities against a space-preference structure. A model of spatial choice developed within this framework would finally accomplish 'the analytical separation of preference and opportunity,'[25] such as was achieved in the economist's model of choice many decades ago. The theory and methods of scaling applied to paired-comparison data provide an appropriate analytical structure for solving many geographical problems in which spatial choice is present.

Notes

1 P. R. Gould (1967) 'Structuring information on spacio-temporal preferences,' *Journal of Regional Science*, 7 (supplement), 259–74; P. R. Gould (1973) 'On mental maps,' in R. M. Downs and D. Stea (eds) *Image and Environment*, Chicago, Aldine.

2 G. L. Peterson (1967) 'A model of preference: quantitative analysis of the perception of the visual appearance of residential neighborhoods,' *Journal of Regional Science*, 7, 19–32.

3 W. Christaller (1966) *Central Places in Southern Germany*, Englewood Cliffs, N.J., Prentice-Hall. Others have drawn attention to this postulate and to its critical role in central place theory: B. J. L. Berry and W. Garrison (1958) 'Recent developments in central place theory,' *Papers of the Regional Science Association*, 4, 107–20; G. Rushton, R. G. Golledge and W. A. V. Clark (1967) 'Formulation and test of a normative model for the spatial allocation of grocery expenditures by a dispersed population,' *Annals of the Association of American Geographers*, 57, 390; G. Rushton (1969) 'Analysis of spatial behavior by revealed space preference,' *Annals of the Association of American Geographers*, 59.

4 C. H. Coombs (1964) *A Theory of Data*, New York, Wiley, especially Part 4, 'Stimuli comparison data'; H. A. David (1963) *The Method of Paired Comparisons*, New York, Hafner; J. P. Guilford (1954) *Psychometric Methods*, New York, McGraw-Hill, Chapter 7, 'The method of pair comparisons' and Chapter 8, 'The method of rank order'; H. Gulliksen (1946) 'Paired comparisons and the logic of measurement,' *Psychological Review*, 3, 199–213; J. B. Kruskal (1964) 'Multidimensional scaling by optimizing goodness to fit to a nonmetric hypothesis,' *Psychometrika*, 29, 1–27; J. B. Kruskal (1964) 'Nonmetric multidimensional scaling: a numerical method,' *Psychometrika*, 29, 115–29; R. N. Shepard (1962) 'The analysis of proximities: multidimensional scaling with an unknown distance function,' *Psychometrika*, 27, 125–39; L. L. Thurstone (1927) 'A law of comparative judgement,' *Psychological Review*, 34, 273–86; W. S. Torgerson (1958) *Theory and Method of Scaling*, New York, Wiley; W. S. Torgerson (1965) 'Multidimensional scaling of similarity,' *Psychometrika*, 30, 379–93.

5 L. Guttman (1946) 'An approach for quantifying paired comparisons and rank order,' *Mathematical Statistics*, 17, 144–63.

6 ibid., 145.

7   Kruskal, 'Nonmetric multidimensional scaling' (see note 4 above).

8   This sample of 603 households was taken in the spring of 1961 by the staff of the Statistical Service Division, Iowa State University. The questionnaire was designed by Professors E. Thomas and W. Macki, and the survey was financially supported by the Iowa College-Community Research Center and by the Bureau of Business and Economic Research, University of Iowa. The purpose of the study was to measure the economic impact of the expenditure patterns of the rural populations of the state on towns of various sizes and at various distances, and to gain some insight into the probable effects of continued decrease in the rural population of the state on these types of communities. A description of the sample with maps showing the locations of the respondents can be found in G. Rushton (1966) *Spatial Pattern of Grocery Purchases by the Iowa Rural Population*, University of Iowa, Bureau of Business and Economic Research, Monograph 9, Appendix A, 103–12.

9   The diagonal elements in Table 4.2 become greater than zero whenever a locational type is chosen by a household that has an alternative choice open to it belonging to the same locational type.

10 An empirical study contains some calls for which there are no data, not because the sample size is too small, but rather because in the area where the data were gathered there may be several pairs of locational types which, whenever they occurred together, were always accompanied by a third locational type that was preferable to the other two.

11 Torgerson, 'Multidimensional scaling of similarity,' 380 (see note 4 above); R. N. Shepard (1966) 'Metric structures in ordinal data,' *Journal of Mathematical Psychology*, 3, 287–315.

12 This description of the scaling problem follows closely that in Kruskal, 'Nonmetric multidimensional scaling' (see note 4 above) and Shepard, 'The analysis of proximities' (see note 4 above).

13 R. P. Abelson and J. W. Tukey (1963) 'Efficient utilization of non-numerical information in quantitative analysis: general theory and the case of simple order,' *Mathematical Statistics*, 34, 1347–69; Shepard, 'Metric structures in ordinal data' (see note 11 above).

14 M. G. Kendall (1955) *Rank Correlation Methods*, 2nd edition, New York, Hafner, 156.

15 Coombs, *A Theory of Data*, 353–9 (see note 4 above).

16 ibid., 347.

17 L. Curry (1967) 'Central places in the random spatial economy,' *Journal of Regional Science*, 7 (supplement); R. G. Golledge (1967) 'Conceptualizing the market decision process,' *Journal of Regional Science*, 7 (supplement); R. G. Golledge and L. A. Brown (1967) 'Search learning and the market decision process,' *Geografiska Annaler*, B, 49(2).

18 Coombs, *A Theory of Data*, 33 (see note 4 above).

19 C. G. Hempel (1965) 'A logical appraisal of operationism,' in *Aspects of Scientific Explanation*, New York, Free Press, Chapter 5.

20　H. S. Houthaker (1961) 'The present state of consumption theory: a survey article,' *Econometrica*, 29(4), 718. See also W. Leontief (1947) 'Introduction to a theory of the internal structure of functional relationships,' *Econometrica*, 15.

21　A. Tversky (1967) 'A general theory of polynomial conjoint measurement,' *Journal of Mathematical Psychology*, 4, 1–20; R. D. Luce and J. Tukey (1964) 'Simultaneous conjoint measurement: a new type of fundamental measurement,' *Journal of Mathematical Psychology*, 1, 1–27.

22　Torgerson, *Theory and Method of Scaling*, 59 (see note 4 above).

23　Guilford, *Psychometric Methods*, 192 (see note 4 above). The ratio N/n indicates the average number of choices per stimulus. In this study N/n was approximately 250.

24　R. N. Shepard (1964) 'On subjectively optimum selection among multi-attribute alternatives,' in M. W. Shelly and G. L. Bryan (eds) *Human Judgements and Optimality*, New York, Wiley, 257–81.

25　T. C. Koopmans, 'On flexibility of future preference,' in Shelly and Bryan (eds) *Human Judgements and Optimality*, 243 (see note 24 above).

PART 2

5 Cognitive mapping: a thematic analysis

Roger M. Downs

Introduction

Re-reading the papers in the original edition of *Behavioral Problems in Geography* is rather like leafing through the pages of the high-school yearbook for the graduating class of '69. Such reminiscing brings back faces, ambitions and expectations. One can see the influences of favorite teachers, like Kenneth Boulding and Kevin Lynch, etched on the ideas that we were struggling to express. The field of cognitive mapping, a term not then in widespread use, had already been singled out as one of those voted most likely to succeed, to go on to bigger and better things.[1] And why not? Who could be so churlish as to disagree with such a judgement? Surely this was a reasonable conclusion to be drawn from the promise that was displayed in an impressive register of first efforts. Peter Gould's widely circulated, inspirational essay on mental maps, Thomas Saarinen's persuasive blend of subtle geographical questions and rigorous psychological method, Reginald Golledge's and Gerard Rushton's provocative explorations of the bases of consumer behavior, all presaged exciting new insights into traditional geographic problems.[2] To be sure, there was an element of a 'new wine in old bottles' character to these and similar efforts. But then such a character was far from undesirable; if behavioral geographers could breathe new life into

I would like to acknowledge the assistance of David Hodge and James Meyer, both of whom commented on a draft of this chapter.

old geographical concerns, then so much the better. This was to be an evolution, albeit a rapid one, and not necessarily a revolution.

Whatever the chosen characterization – revolution, evolution or even devolution – we must come to terms with the subsequent ten years. In so doing, it would be easy to write a 'standard' review of the literature. Such a retrospective review would begin with a search through the now-voluminous card file; Gary Moore, for example, admits to a file of 332 studies on the cognitive representation of large-scale environments alone.[3] In the face of this abundance, we would try to impose some Linnean classification on the themes and approaches. The resultant review would be a mixture of research chronology and biography, spiced with a critical evaluation of the current status of the field.

Such a review seems inappropriate and unappealing for a variety of reasons: there is, of late, a surfeit of essentially classificatory reviews.[4] Moreover, the review would not match the brash and exuberant spirit of the 1969 volume since it would tend to be a boring and possibly misleading recitation of 'achievements.' The force of this latter objection stems from an awkward but unavoidable problem concerning the nature of these achievements. Any retrospective review of attempts to understand cognitive mapping must perforce deal with a substantial body of negative criticism and disquiet, with the 'so what?' comment that has been all too frequently voiced.

In the face of such criticism, the temptation is to respond by setting the record straight, to tidy up the field and then offer either an apologia or, more embarrassingly, a *post mortem*. Having succumbed to this temptation in the past[5] and finding myself succumbing once again, I decided that the first draft of this chapter, rather than the field itself, should be buried.

If such a response is unacceptable, then what are the alternatives? The criticism is too pervasive and persuasive to be ignored. It is clear that the euphoria of the middle 1960s has been replaced by a mood of disquiet and skepticism. Perhaps one of the most significant indications of change came with Yi-Fu Tuan's review of Gould and White's book, *Mental Maps*.[6] Tuan's feeling of disappointment, his characterization of the mapping exercises as games, and his belief that the mental maps were of little use struck home precisely because Tuan was correct in his guess that: 'any aspiring behavioral geographer now regards the ability to do a Gouldian map as an essential part of his training.'[7] Tuan's questioning of the meaning and importance of mental maps, the roots of the field, was elaborated in an essay on the nature of 'Images and mental maps': 'a time comes, however, when a new field must pause in its flight to reconsider the soundness of its foundation and the kinds of questions it asks. Perceptual geography has reached such a stage.'[8] Tuan's appeal was, as always, eloquent, and it was not isolated; Mercer and Powell, Rieser and Graham also offered radical (in the sense of fundamental) criticisms of the field.[9]

The process of reconsideration has culminated in two pieces of work which capture the spirit of uncertainty pervading both cognitive mapping and behavioral geography. These are the essays in Ley and Samuels' edited volume, *Humanistic Geography*, and the dialogue between, on the one hand, Bunting and Guelke and, on the other hand, Rushton, Saarinen and Downs in a 1979 issue of the *Annals of the Association of American Geographers*.[10] In one sense, these works are but another reflection of the discipline-wide debate over the philosophical bases of human geography.[11] In the context of cognitive mapping, we are faced with a series of issues which have been the foci of continuous debate and controversy.

We can capture the essence of this complex debate with two simple questions. Firstly, has cognitive mapping been successful in attaining its goal of understanding the human use of the environment? The consensus seems to be, no. This judgement leads directly to the second question: how can cognitive mapping attain this goal? Again, it is convenient to posit a simple dichotomy of answers. One group, the 'empiricists', would append a 'not yet' to the answer to the first question. A second and less optimistic group, the 'humanists,' would say, 'not in the ways that have been attempted so far,' and they would then call for a careful reconsideration of the goal in order to allow for an alternative mode of understanding.

It is useful to characterize the positions of the empiricist and humanist groups, although we must be conscious of the danger that characterizations can easily become caricatures. Empiricists are grounded in the positivist tradition, emphasize the scientific method, believe that the subjective world in the head is a reflection and distortion of the objective real world, see place and space as objective, geometrical constructs, and see the mind as brain in a psychological sense. The humanist position is an extended essay in contrasts. Humanists proceed by description, by literal reconstruction, to reveal intuitively *self*-evident essences and meanings. They therefore reject the subjective–objective distinction and emphasize the importance of the world as lived in, the world of personal experience, emotion and value. The mind is best thought of as the human spirit, and thus places are free creations of the human mind: they are centers of emotional meaning. Likewise, space is an emotional surface of values, of likes and dislikes.

It is the recognition of the obvious differences between these two positions and of the role that these differences play in the debate over the value of cognitive mapping that have shaped this chapter. Any attempt to review the current status of cognitive mapping must of necessity address the empiricist–humanist controversy. I have attempted to do this by capturing not so much the specifics of the debate itself as the tone of the debate: the context in which it has occurred; the ways in which questions and answers have been asked, and criticisms framed. One emphasis, therefore, is on the intellectual, sociological context of the last ten years. (Such an emphasis, stemming from the sociology of science,[12] is rare in geography, although Taylor's paper on the quantitative revolution adopts a similar stance,[13] while the 1979 issue of the *Annals of the Association of American Geographers*, reflecting on 'Seventy-Five Years of American Geography,' provides the raw material necessary for a sociological analysis.)

The second emphasis of this review directs attention away from a static inventory of what is known towards an appreciation of the process of knowing, away from achievements to achieving. This crucial shift of attention has a double significance. Firstly, it embodies an emerging consensus about our object of study: we are concerned with environmental knowing, with representing more than representation, with mapping more than maps.[14] Secondly, it reminds us that the process of knowing is hopelessly reflexive: the process is simultaneously the object of study *and* the only means by which that object of study is understood.

If the first emphasis of this chapter is the sociology of knowledge, then the second might be called the psychology of knowledge. Together they permit us to understand the relationships between behavioral geography and cognitive mapping. I will approach these relationships in three ways. Firstly, we must go back to the beginning: Why the concern with environmental knowing? What are our goals? Why did we choose them? How have we pursued them? Secondly, I want to explore the expression of this concern by tracing three basic themes that have dominated research and criticism in the last ten years. The three themes are themselves questions: What is a cognitive map? How can we measure a cognitive map? Why do cognitive maps exist and where do they come from? The answers to these questions are crucially dependent on the 'language' that we use; by language, I mean the words, expressions and metaphors that express our thoughts. Finally, I want to assess the importance of work in cognitive mapping. How far have ideas about cognitive mapping penetrated into the consciousness of human geographers? How can we measure the 'success' of these ideas? How can we resolve the problem of 'language,' a problem that I see as central to the future of the field?

Behavioral geography and cognitive mapping

Nothing more vividly expresses the peculiar status of our object of study than the uncertainty over what to call it. After nearly two decades of work, we still lack a publicly recognizable corporate identity. Names can be generated readily by pairing an adjective with a noun. A representative list of adjectives includes environmental, spatial, mental, cognitive, psycho- and perceptual; possible mates range from cognition, perception, knowing, images, maps and mapping, through to geography itself.

Despite the awkwardness that this nomenclatural confusion causes for compilers of bibliographies and key-word indexes, most people now have a good intuitive sense of what we are pursuing. In simplest terms, our quarry is the 'world in the head.' Ironically, it is when we try to go beyond this simple, intuitively obvious metaphor that we run into problems. These interlinked problems, which continue to haunt us, are false precision, exclusion and intolerance.

I have yet to, nor do I ever expect to, encounter a precise, unambiguous, catholic definition of the field of cognitive mapping. Most of the current definitions suffer from one of two fraternal sins; some suffer from both. The first sin (of hubris?) is the heroic, but ultimately futile attempt at producing an overly fine distinction. Where does perception 'end' and cognition 'begin'? At what spatial scale does the geographer's 'legitimate' interest 'begin' – a room, a building, without the 'direct' range of vision? Where does our disciplinary interest 'stop' – with cognitive processes, with neuro-psychology, with neurophysiology?

The second sin, exclusion, leads us to deny legitimacy to certain ideas and approaches, and thus to excommunicate people. This schismatic tendency underlies the debate over the 'contributions' of empiricist versus humanist approaches to understanding the world in the head, over the appropriateness, for example, of the work of Golledge and Gould versus that of Lowenthal and Tuan. It tempts us to draw lines that, on later reflection, turn out to be questionable if not indefensible. For example, it is generally accepted that geographers are not and should not be 'interested' in brain physiology and neuro-anatomy. But then, how are we to cope with some of the recent implications of neuropsychology, especially that speculative work that suggests a physiological locus for space and spatial thinking?

It may only be another reflection of the sins of hubris and exclusion but there is one final reason why our search for identity via definition has not yet been successful. From my experience over the past ten years, it appears that we have all been guilty of a lack of tolerance of the views of others. At times, this intolerance has descended to a questioning of motives and even ability. Such an overly critical stance has manifested itself in a variety of ways: in bibliographies and reviews that do not mention certain sources; in philosophical debates over the 'right' ontology and epistemology; in the use of the 'nothing but-ery' charge as in 'X's work is nothing but . . . .'

In the face of this atmosphere of querulous uncertainty, it is not surprising that we cannot agree on what to call ourselves. It is fortunate, however, that the lack of an identity has not deterred us in the pursuit of the world in the head.

In their original programmatic statements about the scope and goals of behavioral geography, Cox and Golledge translated the world-in-the-head metaphor into the idea of studies of the mental storage of information.[15] These studies fell into two types. The first type involved the investigation of those behavioral mechanisms that have spatial correlates, and the second the measurement of attitudes towards environmental stimuli. This classification proved to be temporary and the ensuing ten years of research departed considerably from the typology that Cox and Golledge suggested. It is exciting, however, to find that, in retrospect, the research shows a remarkable degree of coalescence around two basic types of understanding. Interestingly, these follow closely from a typology that Piaget used to characterize the general thrusts in the development of the child's understanding of space.

The first type of understanding concerns *fundamental spatial relations:* that is, the capacity to think spatially, to understand the concept of space in the abstract, and to use a space as a vehicle for structuring knowledge and for solving problems. Space is not taken in the sense of any particular environment or geographic space, but as a model framework for expressing simultaneity and the quality of interrelationship. Command over fundamental spatial relations is intimately associated with the process of visualization as in, for example, the formation, manipulation and rotation of mental images. Without question, this type of understanding falls more centrally within the province of cognitive and developmental psychology. It is, however, of more than passing relevance to behavioral geographers. For example, something as fundamental as the use of maps and models necessarily draws upon the ability to comprehend and manipulate spatial relations.

The second type of understanding concerns *environmental cognition*: that is, the comprehension of the arrangement and properties of phenomena on the earth's surface. It is this latter which fits closely with the two research types outlined by Cox and Golledge; it is a literal interpretation of the world-in-the-head metaphor. Geography, together with architecture and planning, has been in the vanguard of work on environmental cognition. It has led us into an almost unbelievable range of areas: we have considered the cognitive version of the environment as a geometrical construct with metric and orientation properties; we have classified types of significant cognitive phenomena (as in Lynch's typology of edges, nodes, paths, districts and landmarks);[16] we have treated the environment as an experiential, emotive construct (the place–space contrast, and the ideas of home and neighborhood); and we have tried to express preferences for various parts of the environment (as in Peter Gould's mental maps).[17]

We cannot make sense of nor evaluate this bipolar body of work without going back to the original motivations for undertaking the research. Why was the world in the head considered to be of such fundamental importance? Any answer to this question is necessarily an exercise in *post hoc* rationalizing. As Abraham Kaplan has argued,[18] there is a considerable difference between the *recon-structed logic* and the *logic in use*, between the ways in which we report what we did and why we did it versus the ways in which things actually occurred. With this proviso in mind, it appears that there are two distinct sets of reasons that account for the emphasis on the world in the head. The first set of reasons stem from its conceptual role as an explanatory construct, and the second from the social structure and social dynamics of geography as a discipline. The distinction between the two is an interesting contrast between the overt and the covert, the rational and the emotional, and the logical and the psychological forces that underpin individual decisions about what to study and how to study it.

As an explanatory construct, the world in the head was central to behavioral geography itself and shared its founding rationale. Cox and Golledge stated in what served as the manifesto for the field that: 'All papers, therefore, are united by a common concern for the building of geographic theory on the basis of postulates regarding human behavior.'[19] The world in the head stood alongside such behavioral concepts as decision models, interaction models, diffusion models and learning theory. Each was to contribute to the belief that: '(i)n order to understand spatial structure, therefore, we must know something of the antecedent decisions and behaviors which arrange phenomena over space.'[20] Each behavioral concept could be brought to bear on a wide range of phenomena that had defied successful explanation: interregional migration flows, intra-urban migration, consumer behavior on the microscale, travel behavior and route selection. The list was only limited by the imagination, courage and audacity of the behavioral geographer. But, as I argued in the introduction, the initial mood of enthusiasm and tolerance was gradually replaced by a significant chorus of criticism. The genuine belief in the value of the behavioral approach as a source of explanatory power gave way to a sense that it was a mere placebo, and then, more disturbingly, to the view that this was not just a harmless panacea but a potentially dangerous nostrum.[21]

To put such a change of mood into context requires that we consider the second reason for an emphasis on the world in the head. If we can accept that the process of knowing is reflexive, that it is both object and means, then we must consider the character of the 'knowers' and the intellectual context within which they operate. In this instance, it demands that we see the emphasis on the world in the head (and on behavioral geography) in the light of the social structure and social dynamics of the discipline itself. By characterizing the social structure of the discipline as covert, I do not mean to suggest that it is deliberately disguised and hidden. I do believe that there is, in general, a conspiracy of silence simply because disciplinary structure is something that we prefer not to discuss openly, at least not in the professional literature. It is an emotionally charged issue that can only be understood as a psychologic.

Many of the structural characteristics of human geography in the late 1960s made the uncritical acceptance of the idea of the world in the head both possible and likely. Moreover, the early workers proved to be very effective in adapting to the opportunities that were available. For example, the timing of the behavioral push was right. It appeared when the campaign to make geography into a social and behavioral science was at its height. Here was an ideal opportunity to demonstrate the meaning and power of such a view of geography, especially since the world in the head led us directly towards the pre-eminent behavioral science, psychology. Geography was also growing rapidly and the developing field of cognitive mapping provided yet another patch of fertile ground for the increasing numbers of professional geographers. It offered niches for faculty and graduate students, niches that were virgin territory for trying out new skills and techniques and which would generate publishable papers and dissertations. There was also a certain undeniable power to the ideas that were being expressed. Like noses and opinions, we all had them; we all had images and mental maps. We could produce them just as readily as the students who served as subjects in nearly all of the early work. The ideas were neat, interesting, intuitively plausible, and self-evident in the best sense of that phrase. They were also familiar and comfortable in that they tapped some of the fundamental well-springs of the discipline. By talking about mental maps and cognitive maps, we were legitimate heirs to the cartographic tradition that Carl Sauer had stated so forcefully:

Show me a geographer who does not need them [maps] constantly and want them about him, and I shall have my doubts as to
whether he has made the right choice of life . . . . Maps break

down our inhibitions, stimulate our glands, stir our imagination, loosen our tongues. The map speaks across the barriers of language; it is sometimes claimed as the language of geography. The conveying of ideas by means of maps is attributed to us as our common vocation and passion.[22]

Inasmuch as the work in environmental cognition involved 'real-world' studies, we were pursuing that equally strong geographic tradition, fieldwork. And J. K. Wright's Latin tag, *terra incognita*, could be used to draw us close to the light of classical scholarship. We were equally fortunate in the power and persuasiveness of some of the early proponents of work on the world in the head: Peter Gould brought a flair for metaphor, for sparkling and muscular language; Yi-Fu Tuan provided an air of calm and detachment that led to penetrating observations about the apparently common-place; David Lowenthal was the widely read, classical scholar who opened up fascinating bodies of literature; David Stea tempted us with concepts and methods from psychology, and showed us how easy it was to run pilot studies.

The cumulative effect of all of these structural characteristics was to lead to a willing suspension of disbelief, to a short period when it was sufficient to be doing something. The thinking would come later, but, when it did come, so did the mounting criticism that was outlined in the introduction. But before we try to consider in detail the sources and causes of that criticism, it is worth trying to disentangle the relationship between cognitive mapping and behavioral geography. Although the former pre-existed the latter by a considerable period of time, by the late 1960s they became inextricably interwoven. It is virtually impossible to make a dis-tinction between who is a behavioral geographer and who a perceptual geographer. The awkwardness of this ill-considered relationship gradually became apparent. This can be illustrated in two ways. Papers began to appear with hybrid titles such as behavioral *and* perception geography. And, to the extent that behavioral geography was pursuing a route that we can crudely label as positivistic, reductionist and model building, some bodies of perception work did not fit into the chosen mold. What, for example, was the relevance of the hazard studies of the Chicago school? (It is interesting to note that this classification problem had already occurred in Cox and Golledge's introduction to the field.)[23]

Even more problematic was the status of the literary and humanistic concerns exemplified by Lowenthal and Tuan.

A second sign of awkwardness remains with us and still poses problems. Geography has gotten itself into a terminological muddle with use of the label, 'behavioral geography.' It is a muddle that has particularly amusing consequences for work on fundamental spatial relations and environmental cognition. Inasmuch as such work is seen as part of behavioral geography, we have to be very careful with the connotations of 'behavioral.' It is so easy to slip from the idea of 'behavioral' to 'behaviorist' and then slide on to 'behaviorism.' The linguistic resonances that are induced are cognitively dissonant. In its most extreme form, behaviorism treats the mind as irrelevant and anything mental as a frivolous epiphenomenon. To a radical behaviorist, consciousness and mind are empty words, mere semantic puffs devoid of empirical meaning, that have been invented to provide intuitively satisfying but spurious explanations. And unfortunately, work on the world in the head has come under attack on precisely these grounds. (For an instance of such an attack, see Bunting and Guelke; Richards offers an interesting counter statement. The overall validity of such attacks will be considered in the next major section of this chapter.)[24]

This awkward relationship became exposed as the criticism of both behavioral geography and cognitive mapping grew during the middle 1970s. Much of this criticism arose because of the way in which the fields developed and particularly because of the way in which research was conducted and reported. Much of the work seemed to be shaped by a preoccupation with technique; thus, for example, there is a persistent fascination with intricacies of multi-dimensional-scaling techniques, with the 'progression' from MDSCAL through INDSCAL to the latest program appropriately named ALSCALE. Such technical fascination appears incongruous in the face of a lack of concern with data that are reliable and valid. To many people, studies appeared to be 'hit and run' in character. Ironically, the apparent lack of cumulative, coherent effort is accentuated by the inevitable centrifugal character of behavioral geography. As part of its chosen mandate, behavioral geography was to search far and wide beyond the conventional bounds of geography for suitable methods, techniques and concepts.

How can we understand this shift from enthusiasm to disdain, from statements of great expectations to accusations of lost opportunities? What is the current status of work on cognitive mapping and what is its relationship to the original goals of behavioral geography? We can only answer these questions if we consider the thematic character of cognitive research in more detail. In the preceding paragraph, I deliberately couched my statement of the criticism in the conditional tense. I did so because I feel that whether or not these criticisms are valid is less important than the question, what is the basis for studying the world in the head?

Cognitive mapping after ten years of research

The most useful way of considering the field of cognitive-mapping research is to identify what seem to be the principal organizing themes. For the researchers in the field itself, these themes serve as the basis for the work that is being undertaken and its guide. For the critics of the field, the themes serve as the basis for what are essentially negative and destructive comments that deny the legitimacy of the work.

The three themes are:

1 What is a cognitive map? (The question applies equally well to any of the synonyms for the fundamental concept: mental map, image, schema or representation.)
2 How can you measure a cognitive map? Or, as some would phrase the question, can you measure it? Should you measure it?
3 Why do cognitive maps exist and where do they come from? (This latter question is an awkward amalgam of debate over the functions and origins (genesis) of cognitive maps.)

At present, there is considerable disagreement, confusion and criticism surrounding all three of these questions. It would not be too extreme to suggest that the confusion is sufficient to throw the status of the whole field into question.

I want to argue that, although much of the criticism of cognitive mapping is legitimate and well taken, and although the disagreement and confusion is real and not apparent, many of the problems stem from some fundamental misunderstandings, misunderstandings that were once inevitable but which can be avoided in the future. In particular, the misunderstandings arise from one central cause, that of *language*. By language I mean the extent to which language, that is, the choice of particular words, expressions and phrases, has been structuring, and hence substituting for, thought. This is the obverse of what should occur; that is, thought should structure and dictate the choice of language. For example, behavioral geographers have spoken of maps without thinking through the profound connotations and implications of that word. Maps are 'things' that we 'have,' an expression that has been responsible for a considerable amount of grief and confusion. From these inauspicious starting points, we arrive at the position that if maps are things that we have, then they must 'have' some material existence somewhere; this immediately takes us into the perilous area of the debate over the relation between mind and body. And, if we are not careful, we find ourselves smuggling in something (or some-one) resembling a geographic homonculus because, if we have maps, then who is doing the map reading, especially when 'we' ourselves are not conscious of doing it? One can demonstrate the same type of linguistic problem via the second of our themes, methodology. For example, if we believe that people 'have' maps and that these maps are stored away in the brain (or mind), then we must try to 'get at' them. Thus we find ourselves speaking of 'extracting' maps, of 'unpacking the contents of the mind', of performing some form of exploratory surgery to get at them. But perhaps people do not 'have *any*thing' stored; perhaps they can respond to our demands and generate '*some*thing' for us. Or perhaps they do have something but that something is not a map as a geographer conceives a map to be. In this case, our questions would be totally inappropriate. And this is written only from an empiricist perspective; a humanist would not be surprised at such methodological confusion. The humanist would not presume to bring such preconceived constructs as maps to a 'subject.' The humanist would not be on the 'outside looking in' at 'things.'

Without question, the misunderstandings are often the result of eagerness and impatience, of innocence and naivety. But we have failed to appreciate the extent to which we have been carried along by the sheer force of our expressions. The consequences of this linguistic drift are incredibly important both to understanding the past history of the field *and* to its future development. Two consequences seem particularly important in the context of this paper. Firstly, we have failed to maintain a vital distinction between metaphor and analogy; for an elaboration of this argument, see my forthcoming article.[25] In the most simple terms, a metaphor is a figure of speech: it allows us to avoid the freezing constraints of everyday language and cliché, to evoke a sensory image that gives a 'feel' for an idea, and to achieve a powerful expressive, evocative impact. An analogy is an attempt at explanation: it establishes a functional equivalence between some of the properties of a lesser-known and a well-known situation. The metaphor–analogy distinction is between expression and explanation, between image and model, between implicit and explicit understanding. The danger lies in mistaking a metaphor for an analogy, in allowing a powerful means of expression to become thought of as a pseudo-explanation. In our present case, the problem is captured by the title to Graham's paper: 'What is a mental map?' As she argues: 'We understand well what maps are, but what is the force of calling them "mental" maps?'[26] The problem of the metaphor–analogy relation is one that we cannot afford to ignore any longer if we want to build on the idea of a cognitive map. We must decide whether the map is an expressive metaphor or an explanatory analogy, whether we can use our understanding of, say, cartography to help illuminate the structure and functions of internal spatial representations. (The basis for this choice will be discussed under the heading, 'What is a cognitive map?')

The second consequence is also a point of serious confusion, and it shares many similarities with the metaphor–analogy relation. It appears that behavioral geographers have failed to appreciate the distinction between models *of* how a process 'actually' works and models that suggest that the process works *as if*. These two statements have profoundly different implications for the types of explanation that one seeks. For example, the Piagetian tradition, in explaining the child's progressive comprehension of fundamental spatial relations, speaks of the child as developing an understanding of Euclidean geometry. The key question is, who 'has' the understanding? Does this mean that the child 'discovers' or is 'taught' Euclidean geometry? Or does it mean that the child's behaviors can be modeled (by outside observers) 'as if' the child possessed an understanding of Euclid? The 'model of' versus the 'model as if' distinction is closely related to the question of the status of concepts that serve as intervening variables. Again, behavioral geographers (and their critics) seem to be confused. Do intervening variables (such as cognitive maps) serve to 'come between' and 'link' two

other concepts or, invoking Occam's scythe, do they unparsimoni-
ously 'get in the way of' the same two concepts?

All of these questions and distinctions stem from the basic issue of
language and its relation to thought. In the remainder of this
section, I want to explore how the language problem has interacted
with the development of the three basic themes. It is my belief that
the interaction has had unfortunate effects on the field of cognitive
mapping and it is these effects that determine our critical judgement
of its current status.

What is a cognitive map?

It is impossible to say who first used the term 'map' to express the
idea of the 'world in the head.' Even if we could trace the lineage of
the idea, we could not necessarily resolve its current, confused
status. Therefore, as a convenient starting point, let us return to
Edward Tolman's classic paper in psychology. And, without any
question, it is clear that the cognitive map was a metaphor *and* an
'as if' statement. Several excerpts establish this point:

> We believe that in the course of learning, something *like* a field
> map of the environment gets established in the rat's brain . . . .
>
> Secondly, we assert that the central office is far more *like* a map
> control room than it is like an old-fashioned telephone
> exchange . . . the incoming impulses are usually worked over and
> elaborated in the central control room into a tentative, cognitive-
> *like* map of the environment.[27]

That the cognitive map remains only a convenient metaphor is clear
from Waldo Tobler's argument that:

> [W]e assume that the subjects being studied have a represen-
> tation of their environment and that this is somehow map-like
> and can be observed by some type of measurement procedure.
> I am not convinced that the basic assumption is meaningful, but
> I have been unable to devise an experiment that would force me
> to give it up. Clearly, some representation of the environment is
> required, but whether this is hierarchical or maplike is not
> known.[28]

We can gain some provocative insights into the generation of
metaphors if we consider Jerome Bruner's suggestions about the

'left- to right-hand shift.'[29] Bruner views his own creative thinking as being based on a metaphoric search for hunches: these hunches are the 'combinatorial products of [his] metaphoric activity.' The metaphors themselves must be tamed or shifted from the left to the right hand, from the intuitive, emotional side to the objective, rational side. In this way, there is progress from metaphoric hunch to testable hypothesis. What makes Bruner's argument even more pertinent to our discussion is his claim that such processes are more evident in psychology because it is a discipline in which the theoretical apparatus is not so well developed that it lends itself to generating hypotheses. Moreover, because psychology is, in Bruner's view, a young, insecure profession, psychologists do not like to profess their humanity and admit that such intuitive searches for metaphor are commonplace.

The parallel with geography is uncanny. Bruner has captured precisely the situation in which we find ourselves. The cognitive map is still basically a metaphoric hunch, one that appeals both emotionally and logically to geographers. It is a 'natural' idea, one that we would be perverse to abandon. That the shift from metaphor to hypothesis and evidence has not yet occurred in a convincing way is undeniable; on the other hand, that the shift cannot occur is a nonsense statement. That shifts do occur is shown, for example, in the current pursuit of Karl Pribram's neural hologram metaphor for brain–memory relations. Just as there *may* be physiological evidence for the 'existence' of neural holograms, so too are there hints that the cognitive map has a physiological existence.[30]

We can see points of agreement even among those who favor alternative metaphors for the world in the head. Whether we use map or representation, we are making the following points:

1 When we speak of the world in the head, we are concerned with the *form of knowledge*;
2 We assume that the constructivist position about the nature of cognition holds; that is, knowledge is acquired over time and space through experience; and
3 That an understanding of the form and genesis of knowledge is *necessary* for a satisfactory understanding of spatial behavior.

We capture all of these points with the map metaphor. Part of

Bruner's taming process depends upon empirical research and thus we must turn to the second of our themes.

How can we measure cognitive maps?

If we are to accept the map as a metaphor, then we must be cautious in its extrapolations to such popular expressions as 'a whole atlas of mental maps,' 'the retina of the mind's eye' and 'the mind's eye visualizes.' What are we implying in these instances about the form of knowledge? What is it that we can measure? Answers to these questions depend upon the distinction between *internal* and *external* forms of knowledge.

Let me begin with the external form of knowledge. From this perspective, the methodological problem is one of finding ways in which people can generate or express knowledge. Any particular mode of expression will generate a map or representation. Much of the confusion and controversy over methodology could be removed if we were to accept the following characterization of the process of externalizing knowledge:

1 People have access to multiple strategies for representing knowledge externally; thus we should expect to make use of a wide range of techniques in our research. There is no incompatibility nor necessary conflict between map sketching, model building, triad distance judgements, verbal descriptions, recognition tasks, adjective checklists and verbal scaling devices.

2 We can view each one of these modes of representation as a 'language' for expressing knowledge. Thus we should expect variations in the number of languages available to a given person, in the 'fluency' with which a particular language is used, and in the preference for given languages when confronted with a specific problem.

3 These modes of representation are not extracting a given and fixed representation from some storage container. They are permitting the expression (read externalization) of knowledge in a given situation and thus we should expect different representations of the same environment at different times.

The implications of such a position are numerous. Purely from a methodological point of view, they force us to reconsider some of the methodological debates that have emerged during the past ten years. In particular, given the view that people are generating an external representation and not extracting some pre-existing map or image, many of the debates turn out to be non-issues.

Two such debates are worth discussing in detail: that over the appropriate philosophical stance and that over the fate of reductionism. Perhaps nothing has attracted more attention than the debate over ontology and epistemology. It had its seeds in the original objectives of the behavioral approach. What is the appropriate philosophical position – positivism, phenomenology, existentialism?

All are being practised in current work on the world in the head. The interaction between them has been, to put it mildly, acerbic. (The character of the debate is best exemplified by the discussions between Bunting and Guelke and their commentators Rushton and Saarinen and Downs, and by Buttimer's remark that phenomen-ology is not a 'panacea for disillusioned positivists.'[31] The tone of the empiricist–humanist controversy had led both Buttimer and Entrikin to appeal for a dialogue and a reconciliation.)[32] A score-keeper for the debate would conclude that positivists are viewed with glacial disdain; some would say that they are passé. Humanists are dividing into existentialists and phenomenologists, with the latter group further breaking down to groups focusing primarily on Ponty or Husserl. Sometimes the philosophical identifications are made voluntarily; others are pinned on the unwilling recipients who are then exhibited as specimens. There is nothing wrong with this philosophical awareness *except* inasmuch as it becomes polemical and divisive.

The stage for division was unfortunately set in the original volume of *Behavioral Problems in Geography*. It called for an attempt to operationalize concepts that had previously been used in a subjective and descriptive manner. As the call for rigor and empirical testing was being answered, the answers were greeted by a vociferous group of critics. It is precisely the efforts to be rigorous and to make tests that have provided the ammunition for those who feel that much of the work of the last ten years has been sterile, mindless and – that most terrifying of all criticisms for a human geographer – *in*humane. The image of white-coated, emotionless phrenologists applying callipers to the mind is an apt caricature of this criticism. One major result of the rigorous, empiricist work has been to foster its mirror image, that is, the encouragement of work in a subjective and descriptive manner. It would be a mistake, however, to see this work as a reaction to the success of rigorous operationalization because, firstly, it has not been a roaring success and, secondly, the subjective–descriptive style predated the behavioral push of the late 1960s.[33] Nevertheless, we are confronted with a confusion of basic approaches; as we have argued elsewhere,[34] the confusion is less of a problem than it appears to be. There are indeed multiple ways of knowing, of arriving at an understanding of the process of environmental knowing. The process of knowing is indeed reflexive.

This paradox of operationalism is matched by the paradox of reductionism. In establishing the scope of behavioral geography, Cox and Golledge pointed to the novelty of the approach in its 'deliberate attempt to unpack and identify these [behavioral] elements, to examine their specific effects on spatial activity.'[35] Unfortunately, the process of disaggregation proved to be more destructive than had been imagined. That which was being unpacked turned out not to be a simple machine; it was not a black box that could be converted into a white box by a process of re-assembly. Metaphors such as 'map reading' and 'map making or map-control rooms' did not help. In a sense, after ten years of disassembling, we are left with a stack of oddly shaped parts. In the case of cognitive mapping for example, we are just beginning to understand cognitive-distance transformations, but we are unsure how such transformations are affected either by the nature of spatial experience or the nature of the environment. Moreover, we do not know how these transformations are related to or used in such basic movement decisions as the journey to shop or to work. The translation from cognition to action remains as far away as it did when Miller, Galanter and Pribram made use of that famous comment by Guthrie about Tolman's rats:

Signs, in Tolman's theory, occasion in the rat *realization*, or *cognition*, or *judgement*, or *hypotheses*, or *abstraction*, but *they do not occasion action*. In his concern with what goes on in the rat's mind, Tolman has neglected to predict what the rat will do. So far as the theory is concerned the rat is left buried in thought; if he gets to the food-box at the end that is his concern, not the concern of the theory.[36]

What we lack is a blueprint for re-assembly. We know a considerable amount about the pieces but we are like 'all of the king's men who couldn't put Humpty Dumpty together again.' Without some overarching, comprehensive theory, we will remain at this level. All of the while we retain our predilection for correlational analysis, we will continue to find correlations between many of the pieces, but the meaning of the pieces will continue to elude us.

Naturally, these problems have been seized upon as yet another sign of impending bankruptcy in both behavioral geography and cognitive mapping.[37] There is an alternative perspective. If we appreciate that the answers we get are a direct function of the ways in which we phrase our questions (the externalization process), and if we are careful to recognize the language problem, we can begin to recover.

We can paint exactly the same picture if we consider the other side of the knowledge question. So far we have begged the questions, what is meant by the internal form of knowledge and what does it mean to have a cognitive map? Although we are still troubled by the nuances of 'have,' it is clear that an answer to the question can follow from one of two interpretations of 'have.' The first interpretation sees 'have' as a psychological statement and the second as a physiological statement. Both interpretations remind us of the difficulty of putting boundaries around the 'legitimate' concern of behavioral geographers as they explore cognitive mapping. Both point to the inherent fascination of the subject and to the deep philosophical issues that we must explore.

The psychological interpretation of 'having' is multifaceted. In brief, we can adopt one of two positions.

1 That what we have is 'formless': this is best expressed by Polanyi's two terms, tacit and explicit knowledge.[38] Explicit knowledge can be represented and externalized; tacit knowledge is nonverbal, nonimageable, inarticulate, acritical (that is, beyond immediate critical inspection). In this conception, what we have is the capacity to convert from tacit to explicit knowledge, from internal to external via the process of cognitive mapping.[39]

2 That the internal form of knowledge exists as systems of propositional statements, as images, or as some complex combination of both.

Note that, in both positions, the constructive view of knowledge holds.

The physiological interpretation of 'having' takes us even further away from the apparent core of the geographer's concern. Again, in brief, there are suggestions that having a cognitive map can be translated into either:

1 Neurophysiological statements: thus the concept of space can be 'localized' as 'existing' in the parietal lobe of the right hemisphere; or

2 Neuropsychological statements: the concept of space depends upon the activation of systems in various parts of the brain but it is not 'in' any one localized part of the brain.[40]

The ways in which we think and talk about the world in the head are inseparable from the ways in which we study that same world. By assuming that, for example, 'map' meant something akin to a cartographic map, we determined ways of gathering, analyzing and interpreting data. The point is not that such assumptions are necessarily wrong; rather, they should be examined with care in order that we can be sure that we are asking the appropriate questions of ourselves and of others. By now it should be clear that the problems of language and metaphor are hiding some fundamental questions. It has taken ten years even to begin to see the questions; the answers are not in sight.

Why do we have cognitive maps and where do they come from?

It is impossible to separate the debate over the functions of cognitive maps from the debate over their origins. In fact, given the functionalist stance that most people adopt,[41] it would make little sense to try to separate the two. By adopting a genetic standpoint (as in Piaget's genetic epistemology), it is possible to account for functions and origins on the basis of three complementary 'time' scales:

1 Phylogenesis: two sources of evidence are brought to bear to provide a speculative account of why and how cognitive mapping evolved as a human adaptation to the demands of the spatial environment. One stream derives from comparative studies of the wayfinding and information-processing systems

of nonhuman species; the second stream is a mixture of human biological evolution, physical anthropology and the history of ecological systems.

2 Ontogenesis: by concentrating upon the life span of a single individual, a parallel account of functions and genesis is possible. Aside from such questionable assertions as the one that ontogeny recapitulates phylogeny, the central organizing ideas are those of Piaget.

3 Microgenesis: this third perspective focuses on time scales within which ontogenesis is assumed to be unimportant. It provides accounts of the interaction between learning and experience.

In general, behavioral geographers are 'consumers' of ideas about the genesis and functions of cognitive mapping. As a consequence, our contributions have been minimal and the 'language' problems are much less significant than they are with respect to the first two themes. The major problems, apart from the complexity and uncertainty of the ideas themselves, center on the adequacy of the arguments (what is an 'acceptable' account?), and the problem of distinguishing between models 'of' versus models 'as if.'

The importance of cognitive mapping

I am sure that everyone is now expecting, as a finale, one of those climactic summaries that says, 'Yes, it was well worth the effort. We now know a lot of valuable things and there is every reason to believe that we will know much more, very soon, especially if *you* do the necessary work.' As Oscar Wilde remarked, good advice is more easily given than taken. We do not need any more exhortations and appeals to others. Nor do we need their opposite. We have had too many 'candid' reviews that are but thinly disguised polemics, the likely outcome, if not purpose, of which is to destroy the field of cognitive mapping. Let me try to reach a temporary assessment by considering three issues: the extent to which ideas about cognitive mapping have penetrated both within and without the discipline; the possible ways in which we can measure the success of those ideas; and the potential for the resolution of the 'language' problem.

Cognitive mapping: from within and without

Perhaps the best way to survey the extent to which ideas of cognitive mapping have penetrated into geographic consciousness is to list a series of indicants. On the subject of cognitive mapping, papers have appeared in all of the major geographic journals, manuscripts are being submitted for publication in geometrically increasing numbers, theses and dissertations form an increasing percentage of those listed in the *Professional Geographer's* annual survey, and symposia are a regular feature at professional meetings. Cognitive mapping or one of its synonyms is a professional specialization in the *Association of American Geographers'* listing, a part of the American Geographical Society's *Current Publications* key-word system, and a requested skill in *Jobs in Geography* advertisements; it is an essential topic in newly written or revised introductory textbooks, and is a part of the British Open University curriculum.

The cognitive-mapping approach has been used in studies spanning the common dimensions of geographic research. Work has covered all time periods, spatial scales and contexts (urban through rural), and has used all approaches (descriptive through analytic). All of these indicants suggest widespread acceptance within the social structure of the discipline.

Of equal significance is the exposure outside the discipline of cognitive-mapping ideas. *Newsweek* (1976) discussed mental maps of Los Angeles and Philadelphia, *Harpers* (1978) had an article on the popular geography of country music, and *New West* (1979) offered readers' mental maps of California. On a more serious level are the increasingly strong ties between geography and psychology. Whereas a few years ago geographers were literally going cap in hand to psychologists for help, the relationship has swung, albeit grudgingly, towards symbiosis. As the psychologists ventured out of the confines of their laboratories into the 'real world,' they realized the need for experienced local guides. And, at least on certain issues (such as the theory of cartography and questions of space), geographers are instructing psychologists.

If we were to rest content with all of these indicants, we would have to conclude that cognitive mapping had diffused rapidly and had become entrenched both within and without the discipline. But there is a nagging feeling that we might be mistaking notoriety for fame. The second part of an assessment demands an evaluation of the claims to fame based on the success of the ideas.

The success of cognitive mapping

There are no universally accepted criteria against which the success of any branch of geography can be measured. Therefore, let me outline three criteria that seem to be important to both cognitive mapping and behavioral geography.

First is the issue of 'gut feeling' or intuitive plausibility or *truth via coherence*: for many people, adequacy and success reduces to a gut feeling of credibility. In the end, 'it' (the idea) makes sense, although the feeling is scarcely one that can be articulated. In this respect, there is an interesting parallel with the 1950s debate in geography over free will versus determinism. Some people argued that deterministic explanations of human behavior could be assailed on two grounds. Firstly, they 'reduced' the individuality of human nature to that of a passive agent reacting with no free will. Secondly, the deterministic argument did not describe what 'we' knew to go on inside of ourselves. It was unrealistic and Procrustean. Both of these grounds serve as reasons for doing *and* as a measure of success for cognitive-mapping studies. These grounds are more appealing to the humanist tradition than they are to the empiricist school who would prefer to measure success on the second criterion.

Second is the percentage of variance explained or *truth via correspondence*; this criterion was at the heart of the behavioral approach as expounded in the 1969 volume. And, ironically, it is a criterion that we have been made to live to regret. We began by emphasizing the need to translate from the cognition of spatial structures, as they existed in the human mind, to the prediction of the activities and movements which produce those same structures. So far, the translation in a predictive sense has not been made. And have we, therefore, failed? This is only true if we take such a limited view that, carried to its logical extreme, would deny success to most of human and physical geography. If we allow others to hold us too literally to the 'percent-variance-explained' criterion, then we allow ourselves to be hoisted with a petard of our own making. But such an outcome would require a view of knowing and under-standing that is both narrow and unacceptable, especially to some of those doing the hoisting. The humanist tradition, for example, would not accept a narrow science-as-positivism view of knowing and understanding.

There are two important points that bear on the 'percent-variance-explained' criterion. Clark reminds us that we should distinguish between what he calls regression versus cognitive models of behavior.[42] In particular, we should distinguish between prediction and explanation; these have different objectives and different approaches. Perhaps percent variance explained is better suited to assessing predictive success while intuitive plausibility considers a different sense of explanation. This suggestion can be supported by returning to Harvey's original observations about the status of behavioral geography.[43] He suggested that three types of models, those of classical location theory, stochastic location theory, and cognitive–behavioral approaches, were complementary, accounting for different aspects and useful for different purposes. There is clear evidence that, in the field of consumer behavior, such complementarity holds. The three levels are represented by Christaller and Losch, Curry and Wilson, and Golledge and Rushton.

Third is aesthetic appeal or *truth via beauty*; this is an even more nebulous set of criteria for measuring success. It bears some resemblance to the criterion of intuitive plausibility, but is best thought of as a judgement of elegance, simplicity (though not only in the sense of parsimony), power and credibility. It is on these grounds that many of the metaphors must stand. We find it useful or powerful or illuminating to argue that someone operates as if . . . . These metaphors are, in Paul Weiss' phrase, 'the fiction[s] of a speculative mind.' They are contingent in the sense that we expect to go beyond them and to replace them, just as we also expect to go beyond any analogy. It is when parsimony replaces simplicity that the behaviorists reject the cognitive map as an unnecessary piece of baggage, a metaphysical conceit.

The resolution of the 'language' problem

In many ways, the future role of cognitive mapping is bound up with the resolution of the 'language' problem. The current set of metaphors and expressions have served their purpose, but they are rapidly becoming dangerously obsolete. As I have tried to suggest in the second part of this chapter, we can distinguish some thematic questions that reflect the fundamental bases of our concern. It is equally clear that we can reach no simple judgement on the work of the last ten years. If, for example, we insist on holding ourselves rigidly accountable to the aims, standards and criteria that were established in the 1969 volume, then we would have to record a negative verdict. If we are willing to see the research as evolving, as adapting, then we can record a judgement of qualified success. My personal belief is that such retrospective judgements, grades if you will, are of little value precisely because the field has achieved too much momentum and institutional support to be stopped, even should anyone have the folly and audacity to try. What is important are the prospects, that hazy sense of where we might want to go, how, and why; the idea of a 'prospective' review is one step in this direction.

Notes

1 I. Burton (1963) 'The quantitative revolution and theoretical geography,' *The Canadian Geographer*, 7, 151–62; D. Lowenthal (ed.) (1967) *Environmental Perception and Behavior*, University of Chicago, Department of Geography, Research Paper 109.
2 P. R. Gould (1973) 'On mental maps,' in R. M. Downs and D. Stea (eds) *Image and Environment*, Chicago, Aldine; T. Saarinen (1966) *Perception of the Drought Hazard on the Great Plains*, University of Chicago, Department of Geography, Research Paper 106; R. G. Golledge (1967) 'Conceptualizing the market decision process,' *Journal of Regional Science*, 7 (supplement), 239–58; G. Rushton (1969) 'Analysis of spatial behavior by revealed space preference,' *Annals of the Association of American Geographers*, 59, 391–400.
3 G. Moore, 'Knowing about environmental knowing: the current state of theory and research on environmental cognition,' *Environment and Behavior*, 11, 33–70.
4 T. Saarinen (1976) *Environmental Planning: Perception and Behavior*, Boston, Houghton Mifflin; D. Canter (1977) *The Psychology of Place*, London, Architectural Press; J. D. Porteous (1977) *Environment and Behavior: Planning and Everyday Urban Life*, Reading, Mass., Addison-Wesley; R. Downs and J. Meyer (1978) 'Geography and the mind: an exploration of perceptual geography,' *American Behavioral Scientist*, 22, 59–77; D. Pocock and R. Hudson (1978) *Images of the Urban Environment*, London, Macmillan.
5 Downs and Meyer, 'Geography and the mind' (see note 4 above); R. Downs (1979) 'Critical appraisal or determined philosophical skepticism?' *Annals of the Association of American Geographers*, 69, 468–71.
6 P. R. Gould and R. White (1974) *Mental Maps*, Harmondsworth,

Penguin Books; Y. Tuan (1974) 'Review of P. R. Gould and R. White, *Mental Maps*,' *Annals of the Association of American Geographers*, 64, 589–91.

7 ibid., 590.

8 Y. Tuan (1975) 'Images and mental maps,' *Annals of the Association of American Geographers*, 65, 213.

9 D. Mercer and J. Powell (1972) *Phenomenology and Other Non-positivistic Approaches in Geography*, Melbourne, Monash University; R. Reiser (1973) 'The territorial illusion and behavioral sink: critical notes on behavioral geography,' *Antipode*, 5, 52–7; E. Graham (1976) 'What is a mental map?' *Area*, 9, 259–62.

10 D. Ley and M. Samuels (eds) (1978) *Humanistic Geography: Prospects and Problems*, Chicago, Maaroufa; T. Bunting and L. Guelke 'Behavioral and perceptual geography: a critical appraisal,' 448–62 and 471–4; G. Rushton, 'On behavioral and perception geography,' 463–4; T. Saarinen, 'Commentary-critique of the Bunting-Guelke paper,' 464–8; and R. M. Downs, 'Critical appraisal or determined philosophical skepticism?' 468–71, all in *Annals of the Association of American Geographers*, 69 (1979).

11 See, D. Gregory (1978) *Ideology, Science and Human Geography*, London, Hutchinson; and S. Gale and G. Olsson (eds) (1979) *Philosophy in Geography*, Dordrecht, Holland, D. Reidel.

12 R. Merton (ed.) (1973) *The Sociology of Science: Theoretical and Empirical Investigations*, Chicago, University of Chicago Press.

13 P. Taylor (1976) 'An interpretation of the quantification debate in British geography,' *Transactions of the Institute of British Geographers*, NS, 1, 129–42.

14 G. Moore and R. Golledge (eds) (1976) *Environmental Knowing: Theories, Research and Methods*, Stroudsburg, Pa., Dowden Hutchinson & Ross; R. Downs and D. Stea (1977) *Maps in Minds*, New York, Harper & Row.

15 K. Cox and R. Golledge (eds) (1969) *Behavioral Problems in Geography: A Symposium*, Evanston, Ill., Northwestern University, Department of Geography, Studies in Geography 17, 2–3.

16 K. Lynch (1960) *The Image of the City*, Cambridge, Mass., MIT Press.

17 Gould, 'On mental maps' (see note 2 above).

18 A. Kaplan (1972) *The Conduct of Inquiry: Methodology for Behavioral Science*, San Francisco, Chandler.

19 Cox and Golledge, *Behavioral Problems in Geography*, 1 (see note 15 above).

20 ibid., 2.

21 Burton, 'The quantitative revolution' (see note 1 above); Bunting and Guelke, 'Behavioral and perceptual geography' (see note 10 above).

22 C. Sauer (1956) 'The education of a geographer,' *Annals of the Association of American Geographers*, 66, 287–99.

23 Cox and Golledge, *Behavioral Problems in Geography*, 3 (see note 15 above).

24 Bunting and Guelke, 'Behavioral and perceptual geography' (see note

10 above); P. Richards (1974) 'Kant's geography and mental maps,' *Transactions of the Institute of British Geographers*, 61, 1–15.

25 R. Downs (forthcoming 1981) 'Maps and mappings as metaphors for spatial representation,' in L. Liben, A. Patterson and N. Newcombe (eds) *Spatial Representation and Behavior across the Life Span*, New York, Academic Press.

26 Graham, 'What is a mental map?' (see note 9 above).

27 E. Tolman (1948) 'On cognitive maps in rats and men,' *The Psychological Review*, 55, 192, emphasis added.

28 W. Tobler (1976) 'The geometry of mental maps,' in R. Golledge and G. Rushton (eds) *Spatial Choice and Spatial Behavior*, Columbus, Ohio State University Press, 70.

29 J. Bruner (1966) *On Knowing: Essays for the Left Hand*, Cambridge, Mass., Harvard University Press.

30 J. O'Keefe and L. Nadel (1978) *The Hippocampus as a Cognitive Map*, Oxford University Press.

31 Bunting and Guelke, 'Behavioral and perceptual geography' (see note 10 above); Rushton, 'On behavioral and perception geography (see note 10 above); Saarinen, 'Commentary-critique of the Bunting-Guelke paper (see note 10 above); Downs, 'Critical appraisal' (see note 10 above); A. Buttimer (1977) 'Comment in reply,' *Annals of the Association of American Geographers*, 66, 183.

32 A. Buttimer (1976) 'Grasping the dynamism of lifeworld,' *Annals of the Association of American Geographers*, 66, 277–92; N. Entrikin (1976) 'Contemporary humanism in geography,' *Annals of the Association of American Geographers*, 66, 615–32.

33 See, Gregory, *Ideology, Science and Human Geography*, (see note 11 above); Ley and Samuels, *Humanistic Geography* (see note 10 above).

34 Downs and Meyer, 'Geography and the mind' (see note 4 above).

35 Cox and Golledge, *Behavioral Problems in Geography*, 1 (see note 4 above).

36 G. Miller, E. Gallanter and K. Pribram (1960) *Plans and the Structure of Behavior*, New York, Holt, Rinehart & Winston.

37 Bunting and Guelke, 'Behavioral and perceptual geography' (see note 10 above).

38 M. Polyani (1964) *Personal Knowledge*, New York, Harper & Row.

39 Downs, 'Maps and mappings' (see note 25 above).

40 O'Keefe and Nadel, *The Hippocampus as a Cognitive Map* (see note 30 above).

41 S. Kaplan (1972) 'The challenge of environmental psychology: a proposal for a new functionalism,' *American Psychologist*, 27, 140–3.

42 W. Clark (1976) 'Technical and substantive approaches to spatial behavior: a commentary,' in R. Golledge and G. Rushton (eds) *Spatial Choice and Spatial Behavior*, Columbus, Ohio State University Press, 303–11.

43 D. Harvey, 'Conceptual and measurement problems in geography,' in Cox and Golledge, *Behavioral Problems in Geography* (see note 4 above).

6 Behavioral approaches to the geographic study of innovation diffusion: problems and prospects

Marilyn A. Brown

Introduction

For the past several decades, geographers have been keenly interested in spatial aspects of the diffusion of innovations. However, the orientation of geographic diffusion research has changed considerably during this period. Some of these changes have been simultaneous with similar transitions in other subfields of geography; some have led or lagged behind like trends in the discipline as a whole.

This paper discusses behavioral approaches to the geographic study of innovation diffusion: how they have evolved within the larger discipline of geography and how they might profitably be refocused. Thus, it is worthwhile to begin by briefly describing what has become known as the 'behavioral approach in geography.' For more lengthy discussions of behavioral geography, see Golledge[1] and Golledge, Brown and Williamson.[2]

Behavioral geography is concerned with the internal human processes that underlie human spatial behavior. Examples of such processes are learning, perception, attitude formation and decision making. Thus, behavioral geography is not concerned primarily with describing patterns of spatial structural aspects of the physical environment. Rather, it seeks to understand the role of the social and psychological processes which mediate the environment–behavior relationship. Thus, it recognizes individual differences in cognitions and relies heavily upon survey data for model verification.

Innovation-diffusion research was prominent in the establishment of the behavioral approach in geography. It was one of the first areas of geographic inquiry to focus upon the role of information in human decision making, generating by the late 1960s a substantial literature on the role of information flows and change agents in the adoption of new ideas, technologies and goods.[3]

Recently the merits of this behavioral thrust have been questioned, and several alternative approaches to innovation diffusion have been suggested. The nature of constraints related to the actions of marketing agencies and institutions have been highlighted, as has the role of the political economy, unequal access to innovations, and differential benefits for early and late adopters. Not only have these new concerns raised a great number of questions concerning decision-making approaches to diffusion, they have also generated much criticism of the behavioral approach in general. Thus, in addition to being a catalyst of behavioral geography, diffusion research has produced several critiques of behavioral approaches that have helped to establish alternative paradigms within the discipline.

This paper begins by discussing the evolution of innovation-diffusion research in geography and the role of this research in the emergence of behavioral geography. It then describes three critiques of the behavioral approach to innovation diffusion that reflect growing concerns within the discipline as a whole: the market and infrastructure model; the development perspective; and the radical critique. It then suggests how the behavioral components of innovation-diffusion theory can be strengthened to better portray the role of cognitive processes and determine the potential extent of explanation of behavioral variables. It concludes by assessing future prospects for behavioral approaches to innovation diffusion.

Innovation-diffusion research and behavioral geography: two related evolutions

Innovation-diffusion research in geography comes, in part, from the research of cultural geographers on the origin and spread of culture traits. Much of this early geographic research focused upon spatial and temporal patterns, examining, for instance, a culture trait's spatial distribution at different times. It tended to be inductive, drawing inferences concerning origins, means and routes of dispersal of innovations within and between regions.[4]

The early work of Torsten Hägerstrand is in this tradition. Hägerstrand focused upon the diffusion of technical innovations such as grazing improvement subsidies, bovine tuberculosis controls, soil mapping, the automobile and the telephone.[5] Through empirical analysis, he identified three regularities of diffusion: the neighborhood effect (i.e. the tendency for diffusion to be strongly influenced by the friction of distance); the hierarchical effect (i.e. the tendency for individuals in large places to adopt earlier than people in places further down the hierarchy); and the logistic effect (i.e. the tendency of the cumulative level of adoption over time to approximate an S-shaped curve). Further, Hägerstrand suggested how these three empirical regularities covary in the various stages of a diffusion process. The elaboration of these diffusion patterns, and the neglect of generative processes by which they come about, was characteristic of this early research. In fact, the study of diffusion *patterns*, to the neglect of diffusion *processes*, characterizes much subsequent research.[6]

Hägerstrand, however, shifted his emphasis in 1953 with the writing of his doctoral dissertation, *Innovation Diffusion as a Spatial Process* (translated into English by Allan Pred in 1967). He conceptualized adoption of technical innovations as the result of a learning process, the critical change mechanism being information flows through which resistance to adoption is overcome. The underlying assumption was 'that every new bit of information containing reports of new acceptances or reminders about already known acceptances must involve additional pressure and thereby move the recipient further in the direction of acceptance.'[7] Thus, processes pertaining to individual differences in *information* and *resistances* were deemed important.

Interpersonal information flows, especially face-to-face communications, were viewed as most effective in the learning process, although Hägerstrand recognized that messages may also emanate from mass media. To model social communications, Hägerstrand developed the idea of private and mean information fields.[8] On the basis of short-distance migration and telephone traffic, Hägerstrand made several assumptions about the spatial structure of these fields: firstly, that 'the average frequency of contact between any two locations is the same for all locations separated by an equal distance' and, secondly, that 'the expected frequency of contact is higher for nearer locations than for more distant ones.'[9] The mean information field was the mechanism by which diffusion impulses were transmitted within Hägerstrand's Monte Carlo simulation model.

Resistance was also viewed by Hägerstrand as a matter of individual differences. Barriers varying 'from rational economic constraints to a more unreasonable aversion to change' exist.[10] There is an 'unevenly distributed willingness or opportunity to accept the innovation.'[11] Concerning 'economic' or 'opportunity' constraints, Hägerstrand states: 'one must not forget that many innovations cannot be accepted until a whole series of other conditions have been adapted to the new phenomenon.'[12] The notion that people vary in their resistance to change is also incorporated in Hägerstrand's simulation model. Adoption units are allowed to differ in their receptiveness to information such that some units adopt after only one message is received, while others require two or more messages prior to adoption.

Hägerstrand's 1953 work can be viewed as behavioral in that it is primarily concerned with the processes by which individuals come to know about and accept innovations, and it recognizes individual differences in these processes. The reliance upon a simulation methodology, however, limited Hägerstrand's ability to test his ideas concerning adoption as a learning process. Hägerstrand did not interview those individuals whose behavior were being modeled to determine the importance and role of information flows. He simply obtained data concerning the geographic distribution of adopters to provide a standard by which the simulated patterns could be judged and the behavioral assumptions 'tested.' The fallacy of such instrumentalism has already been discussed in the geographic literature.[13]

Hägerstrand's pioneering work spurred much research into the spatial structure and role of *information* flows in the diffusion of innovations and in other areas of geographic inquiry. Some of the subsequent diffusion research added greater complexity to Hägerstrand's model by considering local as well as long-distance information, opinion leaders and cliques.[14] However, because these studies used the correspondence between simulated and actual patterns as a test of model validity, they share with Hägerstrand the instrumentalist's problem of being unable to confirm the hypothesized role of the causal mechanisms. Pattern correspondence is a necessary but not a sufficient condition to warrant acceptance of the proposed behavioral processes.

Other diffusion research proposed deterministic models that relied upon information flows as causal variables. For instance, Hudson conceptualized a differential equations model that portrayed changes in adoption rates as a function of interaction with change agents and social groups. His assumption was that 'the relevant components for a diffusion model in a modern social context seem to be information, influence via communication, and the social structure itself.'[15] Thus, he stresses behavioral variables. Yet at no point are his assumptions examined through survey data. Similarly, Pred suggests that the diffusion of innovation up, down and across urban hierarchies is a function of the organizational structure of firms and information circulation.[16] Yet his units of observation are metropolitan areas, not individuals.

Hägerstrand's notion of *resistance* initiated a concern for the nature of constraints and barriers to diffusion. The spatial pattern and permeability of diffusion barriers came under investigation and were classified.[17] For instance, capability and coupling constraints[18] pertaining to time allocation and time–space packing have been defined.[19] 'Innovations affect the use of resources in space and over time. Some innovations serve to increase capacity and release a certain resource, while other innovations require additional inputs of human time, settlement's space (time), water and energy, and so on.'[20] Innovation diffusion changes the constraints and opportunities pertinent to human spatial behavior.

Resistance due to impermeable cognitive constructs is a behavioral aspect of change which has received little attention from geographers. Although Hägerstrand put forth the notion that there is an uneven distribution of receptiveness to change,[21] few geographers have attempted to identify the age, income, activity patterns and past spatial experiences of people with high aversions toward change and risk taking. This is perhaps partly because survey data would be required. While time of adoption has been used to classify people as innovators, early adopters, late adopters or laggards, it is not a good surrogate measure. Innovators and early adopters are not necessarily less resistant to change, in general. They may simply have received more promotional communications as the result of some diffusion agency's market segmentation policy.

Alternatively, they may be located close to an outlet distributing the innovation.

Emerging critiques of behavioral approaches to diffusion

As illustrated above, the behavioral approach to diffusion represented by Hägerstrand's research had great influence upon the geographic study of innovation diffusion. In particular, it brought attention to communication, learning and other cognitive processes that have helped to strengthen the behavioral component of diffusion and other geographic models.

In so doing, however, Hägerstrandian diffusion models contributed to the neglect of processes pertaining to other than behavioral variables. For instance, until the 1970s, geographers tended to ignore the importance of market and infrastructure factors, the impact of adoption upon the use and distribution of resources, and the role of the political economy and the environment of institutions that support and variously sponsor inventions and purvey information about them. These biases spawned various critiques of diffusion theory, three of which are discussed below: the market and infrastructure model; the development perspective; and the radical critique.

The market and infrastructure model

The market and infrastructure model views the behavioral approach as only a partial explanation of diffusion processes. The decision of the individual to adopt an innovation is seen as only the final step in a series of stages through which the innovation is made available to a population. One must consider supply as well as demand factors. 'In considering both, diffusion of innovation is no longer simply a consumer behavior phenomenon. It is instead a much broader topic requiring consideration of institutional behaviors, by public and private entities, which affect the individual's or household's access to the innovation.'[22]

Lawrence A. Brown is primarily responsible for developing this broad and well-articulated conceptualization of diffusion.[23] A central element of this model is the diffusion agency, the public or private sector entity through which an innovation is distributed or made available to the population at large. These entities are defined broadly to include retail and wholesale outlets, government agencies and nonprofit organizations. As such, the model is meant to apply to diffusion processes in a broad range of contexts.

The establishment of diffusion agencies is considered the first stage in the diffusion process. These agencies may be established by a single propagator (or several working together) as in a multiple-facility location problem, or they may be established by many propagators working independently. They may have existed prior to the diffusion process, or they may be newly established for the innovation. The resulting pattern of diffusion is hypothesized to differ accordingly.

In the second stage, each diffusion agency conceives and implements a strategy to promote adoption among the population in its service or market area. The implementation of these strategies is elaborated in terms of activities including infrastructure provision (institutional and other structures which enable or enhance the use of an innovation), pricing (its level and spatial variation), promotional communications (which vary according to source, channel, content and receiver), and market selection and segmentation (differing market strategies for different market segments). These activities are seen as having predictable spatial impacts upon diffusion patterns.

Market and infrastructure aspects of innovation diffusion have been theoretically and empirically examined by many geographers. They have been shown to be of importance in the diffusion of a variety of profit and nonprofit innovations including Planned Parenthood affiliates,[24] cable television,[25] Montessori Schools,[26] agricultural products and methods,[27] credit cards[28] and Friendly Ice Cream Shops.[29] In total, this literature has extensively documented and elaborated upon the impact of promotional activities undertaken by diffusion agencies, and the role of various other constraints and incentives to adoption.

Nevertheless, it may be argued that the recent emphasis upon market and infrastructural factors has frequently been at the expense of 'behavioral' considerations. Market and infrastructure variables have been shown to correlate with spatial and temporal patterns of adoption, and this correlation is interpreted as evidence of the impact of market and infrastructure factors. Thus, the frequent warning that one cannot deduce process from pattern goes unheeded. Few have elaborated upon the link to patterns in terms of cognitive variables as evidenced by the neglect of survey research on the part of geographers employing this model. As a result, the importance of market and infrastructure factors *in the individual's adoption decision* is not well documented.

L. A. Brown, M. A. Brown and C. S. Craig have modeled the interface between diffusion-agency actions and adoption using cognitive concepts (Figure 6.1).[30] The activities of the diffusion agency are seen as influencing the subjective attributes of an innovation, and therefore adoption, both by altering objective attributes such as pricing and infrastructures, and by disseminating

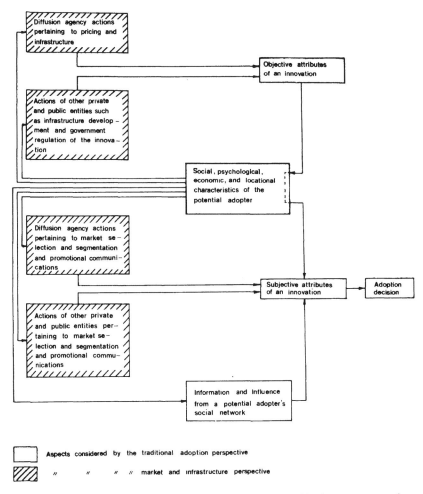

Figure 6.1 A portrayal of the interface between diffusion agency actions and adoption behavior

promotional communications which primarily affect the potential adopter's beliefs about, and evaluation of, these attributes.

Figure 6.1 also identifies the traditional adoption perspective represented by Hägerstrand's research, which is seen as concentrating almost exclusively upon the right, unshaded portion of the diagram. This portion relates adoption to social influence, objective and subjective attributes of an innovation, and the social, psychological, economic and locational characteristics of the potential adopter. In contrast, the market and infrastructure model relates adoption to the actions of diffusion agencies and other relevant public and private institutions, frequently ignoring cognitive aspects of the adoption decision.

In conclusion, the market and infrastructure model has contributed greatly toward the placement of innovation diffusion within the context of diffusion-agency actions. An important shortcoming of the approach is that it focuses on contextual factors at the expense of behavioral ones.

The development perspective

The development critique of innovation diffusion argues that behavioral geographers have neglected issues concerning the impact of diffusion upon individual and societal welfare. It stresses that innovation diffusion is not always beneficial, and is typically more beneficial to some than others. Thus, public policy should not always be designed to bring about adoption most efficiently. Such concerns are frequently expressed with respect to technological change in developing nations. There it has been noted that the diffusion of technological innovations frequently increases regional inequalities and widens the disparities between classes.[31] It has been suggested that 'diffusion waves of development are simultaneously [nondiffusion] waves increasing the relative levels of underdevelopment.'[32]

These realizations undermine the assumptions of much early literature concerning the role of diffusion in social change. Typical of this early literature is Rogers and Shoemaker's discussion of innovation diffusion as the process whereby traditional societies become more modern, change oriented, scientific, rational and cosmopolite.[33] The development perspective also leads one to question the view of adoption as a learning process in which individuals gain information predominantly through social communications. As such, it underlines the significance of the market and infrastructure model.

Innovations have been classified variously as continuous, dynamically continuous and discontinuous,[34] propagator supported or not,[35] consumer or firm oriented,[36] and time saving or time demanding.[37] Only with the recent emergence of the development critique have classifications been based on the equity of their impacts.

In the context of innovation diffusion in a developing country, Yapa has classified innovations according to which factors of production benefit from the diffusion. For instance, labor-augmenting innovations raise the marginal productivity of labor relative to that of capital, thereby favoring peasants; while material-augmenting innovations raise the marginal productivity of capital relative to that of labor, thereby favoring landlords. Since access to the factors of production varies by social class, a factor bias is simultaneously a social bias.[38] Other geographers have studied the differential costs and benefits arising from diffusion according to time of adoption. Berry[39] and L. A. Brown,[40] for instance, argue that 'adoption rent' frequently depends upon time of adoption such that early adopters receive excess profits or greater utility from the innovation. This would occur, for example, when an innovation decreases production costs, but when prices for the resulting product do not drop immediately. Early adopters gain windfall profits while later adopters are faced with decreased prices for their products, such that even with adoption they can only gain enough revenues to remain economically active.

In light of these studies, various public strategies have been criticized, including use of the two-step flow of information by the US Cooperative Extension Service.[41] The two-step flow strategy seeks first to contact 'progressive' individuals who will act as opinion leaders and thereby persuade others to adopt. Income or size of economic operation (e.g. farm size) is frequently used as a surrogate for progressiveness when using this approach. Thus, those individuals who are already prospering most are given the advantage of early information about new techniques, products and other innovations. When adoption rent varies by time of adoption, use of this strategy has particularly perverse effects.

In common with the thrust of the market and infrastructure

model, the development critique stresses constraints to adoption, particularly relative to differential access to resources such as information, means of production and public goods. Yapa and Mayfield, for instance, assumed that adoption requires personal attributes favorable to innovation, sufficient information, and possession of the economic means to acquire the innovation.[42] In an attempt to assess the relative importance of these various factors, they studied the role of three types of variables in the adoption of agricultural innovations by Indian farmers. Biographical data were able correctly to classify 60 percent of subjects as adopters or non-adopters, communications data could classify 64 percent properly, and resources data could classify 74 percent correctly, suggesting that access to resources is more closely related to adoption behavior than communications.

This and other studies employing a development perspective to innovation diffusion have provided substantial evidence that an approach to diffusion in the Third World that employs only behavioral variables is inadequate. Technological change is more than simply a process of information dispersal whereby traditional societies are modernized. Access to resources affects access to innovations; thus behavioral variables may provide little insight into certain diffusions.

One failure of the development perspective is its lack of insight into the relative importance of behavioral and other variables in different types of diffusions. It has also not explained the sometimes high concomitance of level of information and access to other resources.

The radical critique

The radical critique of behavioral approaches to innovation diffusion argues that an understanding of the social, economic and political conditions that operate independently of the individual must precede an understanding of innovation adoption. The context of commodity and social relations influences the structure of channels of communication, access to resources and, therefore, adoption. The nature of the state and its institutions affects the invention process and has a bearing upon the design and implementation of diffusion strategies. As Reiser has put it, 'Elucidation of personal constructs, preferences or attitudes may well demonstrate subjective consciousness, but it does not take account of the concrete possible alternatives inherent in social and economic conditions.'[43]

Recognition of those social and economic relations that influence diffusion allows the analyst to suggest structural changes that may greatly accelerate a diffusion process. For instance, in studying the spatial diffusion of family planning in India, it has been concluded that the females' ignorance about contraceptive devices stems from their spatial and social isolation. At one level a policy recommendation is to overcome this ignorance by the spatial dispersal of information through local nurses who have the wives' confidence. A more radical program would seek to free the wives from their spatial isolation by ending their lives of subservience to the male.[44]

The radical critique concurs with the development perspective in recognizing that diffusion affects the economic, political and social structure of a system. Thus, it suggests a dialectical approach. 'Innovations by this method are treated as both products of contradiction and causes of new contradictions and the process of adoption as both being structured by a social formation as well as changing it.'[45]

The market and infrastructure, development and radical critiques underscore the need to view adoption as more than simply a behavioral process. We cannot understand diffusion outside the context of the social and political economy. Public and private entities can significantly enable, induce or inhibit adoption. Institutional and other constraints to innovation are important, as are differential access to resources and unequal benefits from adoption. Thus, innovation-diffusion research must be broadened through the inclusion of a variety of variables, behavioral and otherwise.

Improving the behavioral component of diffusion research

The communications approach to innovation adoption was portrayed earlier in this paper as dominating much of geographic diffusion research. The various critiques of information-based theories discussed thus far illustrate one aspect of their incompleteness: their failure to consider processes related to markets, institutions and other 'external' or 'contextural' variables. By emphasizing information flows, the communications approach also neglects many aspects of human cognitive processes that might be useful in fusing behavioral or internal with contextual or external variables. These include affective or evaluative processes such as the development of attitudes, the integration of information into evaluative judgements, and the role of personal constructs in human decision making. These psychological aspects are three among many that have received little attention by geographers studying the diffusion of innovations. Each of these approaches is discussed below in terms of how it might provide further insight into innovation diffusion.

Attitude theory

Attitude theory has been employed for many years by consumer behaviorists in the study of new-product marketing. It has only recently been used by geographers to study innovation diffusion. Attitude theory suggests that innovation-adoption behavior is affected by a person's beliefs about the attributes associated with the behavior, and by her or his evaluation of these subjective attributes.[46] For instance, an individual's attitude toward adopting an innovation is related to his or her beliefs about attributes of the innovation, and the utility or value of these subjective characteristics to the individual. When combined with attitudes toward complements of the innovation (i.e. other components of the 'adoption package') and substitutes for the innovation (e.g. the current consumption pattern), these attitudes presumably lead to behavioral intentions, which in turn translate into behavior.

The functional form of the relationship between an attitude and associated beliefs and values has been variously specified.[47] In an application to diffusion, M. A. Brown views an individual's attitude toward adoption of an innovation, A, as a function of his or her attitudes toward various dimensions of beliefs pertaining to adoption, A_j,[48]

In examining attitudes toward innovations, an information dimension would likely be important. Does an individual feel adequately informed concerning the assets and liabilities of the innovation and how to use it? A spatial dimension might be critical. Does the potential adopter feel that the innovation is favorably located in terms of availability or possible delivery services? A third cluster of beliefs might pertain to aspatial attributes of the innovation such as complexity and durability. A personal structural dimension might also be critical. Does the individual feel she/he can afford the innovation? Would the innovation be useful given the individual's particular age, sex, farm size, production practices, etc,? Additional dimensions might be needed fully to represent an individual's global attitude toward adoption of any specific innovation.

In mathematical terms, then:

$$A = f[\{A_j\}, j = 1, \ldots, n]$$

where

$$A_j = \sum_{i=1}^{nj} S_i V_i$$

S_i = the salience or importance of belief i in the individual's adoption decision;

V_i = the individual's evaluation of belief i;

n_j = the number of beliefs associated with dimension j of an innovation; and

n = the number of dimensions of beliefs associated with adoption.

Since the attributes about which judgements are made may include constraining factors, such as 'availability of credit' or 'distance to the nearest agency marketing the innovation,' attitude theory is capable of incorporating many of the market, institutional and infrastructure concerns discussed earlier.

Using the model expressed in the second equation, attitudinal variables have been compared with 'objective' socioeconomic and locational variables, in terms of their ability to discriminate between adopters and non-adopters.[49] The diffusions of five agricultural innovations within Appalachian Ohio were studied. On the average, socioeconomic and locational variables were able properly to classify 62.5 percent of the respondents, while the attitude variables properly identified 73.2 percent of the farmers. Thus, the human decision-making variables were found to provide considerable explanation beyond that provided by variables that are not behavioral.

Information integration theory

Information integration theory also postulates that judgements (and hence behavior) are a function of the individual's evaluation of attributes. However, it does not concentrate upon the role of varying beliefs about attributes of alternatives. Rather, it is concerned with the estimation of algebraic functions that combine evaluations of known attributes into a single judgement. It is assumed that the decision maker identifies relevant attributes upon which to base a judgement, subjectively evaluates alternatives along these dimensions, combines these evaluations in some fashion, and selects the alternative with the most positive overall valence.[50] Thus, adoption is chosen over non-adoption if its attributes combine to form a more favorable judgement.

An impressive array of empirical evidence suggests that a general cognitive algebra underlies human judgement. Support is frequently found for a multiplicative integration principle. This principle is specified in the following equation in terms of m characteristics of behavioral alternative belief:

$$R_k = \prod_{i=1}^{m} X_{ik}^{w_i}$$

where

R_k = the individual's evaluative judgement about alternative k;
X_{ik} = the individual's evaluative judgement about attribute i of alternative k; and
w_i = the importance of attribute i in the overall evaluation of k.

This type of model has been applied by geographers to research on public transportation preference,[51] residential preference,[52] and migration decision making.[53] While it has provided insight into aspects of decision making, particularly regarding the interaction of subjective values, it has not been used to study innovation-adoption processes.

Following the standard experimental design associated with information-integration theory, an application of the theory to consumer goods innovations might consider three attributes of a good such as price, distance to nearest distributor and availability of credit, using three values for each attribute (e.g. prices of $1, $2 and $3). Respondents would be asked to make evaluative judgements concerning the twenty-seven possible combinations of the three attributes. Analysis of variance could be used to assess the role of each attribute both in isolation and in combination with the other attributes, thereby testing the multiplicative nature of the above equation. For instance, is credit viewed as critical only when the innovation is available locally and the price is an acceptable level?

Such an experiment examines probable responses of individuals with uniform information concerning the nature of alternatives. It does not explore the information-acquisition process, which in certain circumstances may explain more of the variation in adoption behavior than does the decision-making process. As such, it ignores the very process that Hägerstrand felt to be most critical in innovation diffusion. However, a comparison of experimentally elicited judgements with actual adoption behavior might identify discrepancies which could lead to a parceling out of behavioral differences due to decision-making processes from those due to information acquisition.

Personal-construct theory

Personal-construct theory assumes that behavior is channelized by the constructs a person sets up to interpret events and thereby predict their future replication. Since individuals differ from each other in these constructions, behaviors also differ.[54] Geographers have elicited and analyzed constructs relevant to urban design[55] and environmental images,[56] but personal-construct theory has not been applied to diffusion processes.

In order to examine the role of personal constructs in innovation adoption, a repertory grid design would traditionally be employed. First the triad method would be used to elicit attributes that distinguish between an innovation and its various substitutes. The individual would then evaluate each alternative according to the elicited attributes. A factoring of the grid of judgements would provide insight into the personal constructs or clusters of attributes that are evaluated similarly, and clusters of alternatives that have similar attributes.

If properly designed, such research could indicate the extent to which various constraints are viewed by potential adopters as important attributes along which items differ. The degree to which similarly constrained population subgroups have similar constructs would provide insight into the impact of the larger-scaled contextual environment upon cognitive processes. Finally, the resulting personal constructs could be used to help formulate attitudinal and information integration studies, both of which require some knowledge of the important dimensions along which alternatives are viewed as differing. The constructs could then be linked to observed behavior, a step that is frequently missing in applications of personal-construct theory.

A prognosis

This paper has argued that the instrumentalist, information-based approach characterizing much past diffusion research is inadequate. Firstly, it has tended to be overly simplistic, viewing innovation diffusion as merely a learning process resulting from social communication. Such a model does not consider individual differences in attitudes, personal constructs or many other psychological processes, and overlooks market and infrastructure factors. Secondly, it compares model-generated patterns with observed patterns as a means of model verification, rather than asking individuals about their attitudes and beliefs concerning adoption.

One promising area of inquiry concerns the relative importance of behavioral versus other processes in different diffusions. Recall the inconsistent findings of M. A. Brown[57] and Yapa and Mayfield[58] in comparing the correlation of different types of variables with adoption behavior. Brown underlined the importance of attitude or choice variables in a case study of agricultural innovations in the US, while Yapa and Mayfield underscored the role of resource or constraint factors in the adoption of agricultural innovations in South Asia. This supports the radical thesis that different causal explanations are required for innovation diffusion within different social, political and economic relations.

It is likely that the importance of behavioral processes also differs across innovations. Processes underlying the purchase of a new high-order good may differ from those underlying the adoption of new convenience products. Differences may exist across types of adoption units, as well. Certainly, where the adoption unit is not an individual, egoistic models of choice behavior will be less important, and perhaps small-group analysis, game theory or other approaches to collective action need to be considered. Community-oriented attitudes rather than the attitudes of individuals may also be useful concepts.

Further, behavioral variables may differ in importance across population subgroups. For instance, Burnett suggests that the set of behavioral alternatives available to the poor is frequently small;[59] thus, delineation of this choice set may provide considerable explanatory power. For wealthier population segments, however, the choice set is typically large, requiring the consideration of attitudes or other choice concepts. Thus, in innovation-diffusion research, constraints may explain the non-adoption of those without access to resources, while choice processes must be considered when understanding the non-adoption of others.

Adopters might be differentiated along a continuum from 'forced' to 'unforced,' or 'involuntary' to 'voluntary.' This continuum reflects some of the same notions behind the dichotomy of voluntary and involuntary migration. Forced adoption might occur when a colonial power imposes a market economy onto a barter one. There, adoption would not necessarily imply 'acceptance' of an innovation. Unforced adoption might be illustrated by choice among new versus old brands of an inexpensive product where behavioral processes would typically be important.

Behavioral research into spatial aspects of innovation diffusion has been a catalyst in the development of behavioral geography. Yet the behavioral component of diffusion theory tends to be highly oversimplified and requires strengthening before its explanatory power can be properly assessed. The recent emergence of radical and other critiques underscores the need to strengthen behavioral components in part by better linking them to broader contextual processes.

Notes

1　R. G. Golledge (1977) 'Behavioral approaches in geography: contents and prospects,' in R. N. Taaffe and J. Odlund (eds) *Geographical Horizons*, Dubuque, Kendall-Hunt.

2　R. G. Golledge, L. A. Brown and F. Williamson (1972) 'Behavioral approaches in geography: an overview,' *The Australian Geographer*, 12, 59–79.

3　T. Hägerstrand (1967) *Innovation Diffusion as a Spatial Process*, University of Chicago Press; L. A. Brown (1968) *Diffusion Processes and Location: A Conceptual Framework and Bibliography*, Philadelphia Regional Science Research Institute; P. R. Gould (1969)

Spatial Diffusion, Washington, D.C., Association of American Geographers, Resource Paper.

4   H. Bobek (1962) 'The main stages in socioeconomic evolution from a geographical point of view,' in R. L. Wagner and M. W. Mikesell (eds) *Readings in Cultural Geography*, University of Chicago Press; D. Stanislawski (1949) 'The origin and spread of the grid-pattern town,' *Geographical Review*, 36, 195–210; C. O. Sauer (1952) *Agricultural Origins and Dispersal*, New York, American Geographical Society.

5   T. Hägerstrand (1952) *The Propagation of Innovation Waves*, Lund, Gleerup.

6   B. Berry (1972) 'Hierarchical diffusion: the basis of developmental filtering and spread in a system of growth centers,' in N. M. Hansen (ed.) *Growth Centers in Regional Economic Development*, New York, Free Press; B. T. Robson (1973) *Urban Growth: An Appraisal*, London, Methuen; D. W. Harvey (1966) 'Geographical process and the analysis of point patterns: testing models of diffusion by quadrat sampling,' *Transactions of the Institute of British Geographers*, 40, 81–95.

7   Hägerstrand, *Innovation Diffusion as a Spatial Process*, 265 (see note 3 above).

8   ibid., 165–241.

9   S. Gale (1972) 'Some formal properties of Hägerstrand's model of spatial interactions,' *Journal of Regional Science*, 12, 199–217.

10 Hägerstrand, *Innovation Diffusion as a Spatial Process*, 264 (see note 3 above).

11 ibid., 148.

12 ibid., 264.

13 J. Agnew (1979) 'Instrumentalism, realism and research on the diffusion of innovation,' *Professional Geographer*, 31, 364–70; D. Amadeo and R. G. Golledge (1975) *An Introduction to Scientific Reasoning in Geography*, New York, Wiley, 228.

14 L. W. Bowden (1965) *Diffusion of the Decision to Irrigate*, University of Chicago, Department of Geography, Discussion Paper; R. P. Misra (1969) 'Monte Carlo simulation of spatial diffusion: rationale and application to the Indian condition,' in *Regional Planning*, University of Mysore Press; G. J. Hanneman, T. W. Carrol, E. M. Rogers, J. D. Stanfield and N. Lin (1969) 'Computer simulation of innovation diffusion in a peasant village,' *American Behavioral Scientist*, 12, 36–45; G. W. Shannon (1970) *Spatial Diffusion of an Innovative Health Care Plan*, Ann Arbor, University of Michigan.

15 J. C. Hudson (1972) *Geographical Diffusion Theory*, Evanston, Ill., Northwestern University, Studies in Geography, 63.

16 A. Pred (1975) 'Diffusion, organizational spatial structure and city-system development,' *Economic Geography*, 51, 252–68.

17 R. S. Yuill (1964) *A Simulation Study of Barrier Effects in Spatial Diffusion Problems*, Evanston, Ill., Northwestern University, Department of Geography, Technical Report; D. F. Marble and J. D. Nystuen (1963) 'An approach to the direct measurement of community

mean-information fields,' *Regional Science Association*, 11, 99–109.

18 T. Hägerstrand (1969) 'What about people in regional science?' *Regional Science Association*, 24, 7–24.

19 T. Carlstein (1978) 'Innovation, time allocation and time–space packing,' in T. Carlstein, D. Parkes and N. Thrift (eds) *Timing Space and Spacing Time*, London, Arnold.

20 ibid., 149.

21 Hägerstrand, *Innovation Diffusion as a Spatial Process* (see note 3 above).

22 L. A. Brown (1981) *Innovation Diffusion: A New Perspective*, London and New York, Methuen, 71–2.

23 L. A. Brown (1968) *Diffusion Dynamics: A Review and Revision of the Quantitative Theory of the Spatial Diffusion of Innovation*, Lund, Gleerup; Brown, *Diffusion Processes and Location* (see note 3 above); L. A. Brown (1975) 'The market and infrastructure context of adoption: a spatial perspective on the diffusion of innovation,' *Economic Geography*, 51, 185–216; L. A. Brown, *Innovation Diffusion* (see note 22 above); L. A. Brown and K. R. Cox (1971) 'Empirical regularities in the diffusion of innovation,' *Annals of the Association of American Geographers*, 61, 551–9.

24 L. A. Brown and S. G. Philliber (1977) 'The diffusion of a population-related innovation: the Planned Parenthood affiliate,' *Social Science Quarterly*, 58, 215–28.

25 L. A. Brown, E. J. Malecki, S. R. Gross, M. N. Shrestha and R. K. Semple (1974) 'The diffusion of cable television in Ohio: a case study of diffusion-agency location processes of the polynuclear type,' *Economic Geography*, 50, 285–99.

26 J. W. Meyer (1975) *Diffusion of an American Montessori Education*, University of Chicago, Department of Geography, Research Paper.

27 M. A. Brown (1977) *The Role of Diffusion Agencies in Innovation Adoption: A Behavioral Approach*, Ohio State University, Department of Geography; R. D. Garst (1974) 'Innovation diffusion among the Gusii of Kenya,' *Economic Geography*, 50, 300–12.

28 M. A. Brown and L. A. Brown (1976) 'The diffusion of BankAmericard in a rural setting: supply and infrastructure considerations,' *Annals of the Association of American Geographers*, 8, 74–8; E. J. Malecki and L. A. Brown (1975) *The Adoption of Credit Card Services by Banks: A Case Study of Diffusion in a Polynuclear Setting with Central Propagator Support*, Ohio State University, Department of Geography.

29 J. W. Meyer and L. A. Brown (1979) 'Diffusion agency establishment: the case of Friendly Ice Cream and public-sector diffusion processes,' *Socio Economic Planning Sciences*, 13, 241–9.

30 L. A. Brown, M. A. Brown and S. Craig (1981) 'Innovation diffusion and entrepreneurial activity in a spatial context: conceptual models and related case studies,' in J. Sheth (ed.) *Research in Marketing, IV*, Greenwich, JAI Press, 69–115.

31 L. S. Yapa (1977) 'The Green Revolution: a diffusion model,' *Annals of*

the Association of American Geographers, 67, 350–9; L. S. Yapa and R. C. Mayfield (1978) 'Non-adoption of innovation: evidence from discrimination analysis,' *Economic Geography*, 54, 145–56; L. A. Brown, R. Schneider, M. E. Harvey and J. B. Riddell (1979) 'Innovation diffusion and development in a Third World setting: the case of the cooperative movement in Sierra Leone,' *Social Science Quarterly*, 60; L. A. Brown, *Innovation Diffusion* (see note 22 above); E. M. Rogers (1976) 'New perspectives on communication and development: an overview,' in E. M. Rogers (ed.) *Communication and Development: Critical Perspectives*, Beverly Hills, Ca., Sage; P. M. Blaikie (1975) *Family Planning in India: Diffusion and Policy*, London, Arnold; P. M. Blaikie (1978) 'The theory of the spatial diffusion of innovations: a spatial cul-de-sac,' *Progress in Human Geography*, 2, 268–95.

32 L. A. Brown (1977) *Diffusion Research in Geography: A Thematic Account*, Ohio State University, Department of Geography, 38.

33 E. M. Rogers and F. F. Shoemaker (1971) *Communication of Innovations: A Cross Cultural Approach*, New York, Free Press, 33.

34 E. Robertson (1971) *Innovative Behavior and Communications*, New York, Holt, Rinehart & Winston.

35 L. A. Brown, 'The market and infrastructure context of adoption' (see note 23 above).

36 E. J. Malecki (1975) 'Innovation diffusion among firms,' Ohio State University, Department of Geography, PhD dissertation.

37 Carlstein, 'Innovation, time allocation and time–space packing' (see note 19 above).

38 Yapa, 'The Green Revolution (see note 31 above); Yapa and Mayfield, 'Non-adoption of innovation (see note 31 above); L. A. Brown, *Innovation Diffusion*, Chapter 8 (see note 22 above).

39 Berry, 'Hierarchical diffusion' (see note 6 above).

40 L. A. Brown, 'The market and infrastructure context of adoption' (see note 23 above); L. A. Brown, *Innovation Diffusion* (see note 22 above).

41 M. A. Brown, G. E. Maxson and L. A. Brown (1977) 'Diffusion-agency strategy and innovation diffusion: a case study of the Eastern Ohio Resource Development Center,' *Regional Science Perspectives*, 7, 1–26.

42 Yapa and Mayfield, 'Non-adoption of innovation' (see note 31 above).

43 R. Reiser (1977) 'The territorial illusion and behavioral sink: critical notes on behavioral geography,' in R. Peet (ed.) *Radical Geography: Alternative Viewpoints on Contemporary Social Issues*, Chicago, Maaroufa (now London and New York, Methuen), 207.

44 Blaikie, 'The theory of the spatial diffusion of innovations,' 287 (see note 31 above).

45 ibid., 279–80.

46 S. Himmelfarb (ed.) (1974) *Readings in Attitude Change*, New York, Wiley; M. Fishbein and I. Ajzen (1975) *Belief, Attitude, Intention and Behavior: An Introduction to Theory and Research*, Reading, Mass., Addison-Wesley.

47   M. Fishbein (1967) 'A behavioral theory approach to the relations between beliefs about an object and the attitude toward the object,' in M. Fishbein (ed.) *Readings in Attitude Theory and Measurement,* New York, Wiley; Fishbein and Ajzen, *Belief, Attitude, Intention and Behavior* (see note 46 above).

48   M. A. Brown, *The Role of Diffusion Agencies* (see note 27 above).

49   M. A. Brown (1980) 'Attitudes and social categories: complementary explanations of innovation adoption,' *Environment and Planning, A,* 12, 175–86.

50   N. H. Anderson (1974) 'Cognitive algebra: information integration theory applied to social attribution,' in L. Berkowitz (ed.) *Advances in Experimental Social Psychology,* 7, 1–101; N. H. Anderson (1974) 'Information-integration theory: a brief survey,' in D. H. Krantz, R. S. Atkinson, R. D. Luce and P. Suppes (eds) *Contemporary Developments in Mathematical Psychology, II,* San Francisco, Ca., W. H. Freeman.

51   J. J. Louviere and K. L. Norman (1977) 'Applications of information-processing theory to the analysis of urban travel demand,' *Environment and Behavior,* 9, 91–106.

52   J. J. Louviere and D. A. Henley (1977) 'Information-integration theory applied to student apartment selection decisions,' *Geographical Analysis,* 9, 130–41.

53   S. Lieber (1979) 'An experimental approach for the migration decision process,' *Tijdschrift voor Economische–Sociale Geografie,* 70, 75–85.

54   P. Kelly (1955) *The Psychology of Personal Constructs,* New York, Norton Press; R. Downs (1976) 'Personal constructions of personal-construct theory,' in R. G. Golledge and G. Moore (eds) *Environmental Knowing,* Stroudsburg, Pa., Dowden, Hutchinson & Ross.

55   B. Honikman, 'Personal-construct theory in the measurement of environmental images: problems and methods,' in Golledge and Moore, ibid.

56   F. Harrison and P. Sarre, 'Personal-construct theory, the repertory grid and environmental cognition,' in Golledge and Moore, ibid.; F. Harrison and P. Sarre (1971) 'Personal-construct theory in the measurement of environmental images: problems and methods,' *Environment and Behavior,* 3, 351–74.

57   M. A. Brown, 'Attitudes and social categories' (see note 49 above).

58   Yapa and Mayfield, 'Non-adoption of innovation' (see note 31 above).

59   P. Burnett (1978) 'Spatial choice versus spatial constraints: alternative approaches to theory with particular reference to travel behavior,' Urbana, paper presented at University of Illinois, March 1.

7 Cognitive behavioral geography and repetitive travel

John S. Pipkin

Introduction

Repetitive travel, including work, shopping, recreational and social trips, is a central theoretical concern in urban and economic geography. Predictive models of travel are also among geography's most distinctive and analytically sophisticated contributions to urban and regional planning.[1] In the pluralistic geography of 1980, there is no consensus that positive, 'objective' description and explanation of such travel should be a principal end or goal of geographic inquiry. Humanists, phenomenologists and others concede the role of repetitive travel in molding perceptions, values and meanings, but demur from the objective, aggregative and mathematical language in which most travel literature is phrased.[2] Radicals, and others who want geography to inform change, question the values concealed in current work,[3] and would regard its products merely as a point of departure for identification and diagnosis of the inequities in opportunities, access and cost that have been well confirmed in the study of urban travel.[4] Notwithstanding these views, the explanation of repetitive travel as conventionally defined remains a principal test of the success of the cognitive–behavioral paradigm in geography. Cognitive–behavioral approaches must produce accounts of travel that are incisive and also distinctive, in that they rest on specifically cognitive or behavioral assumptions. That the paradigm should be judged in these terms is implied by the theoretical and applied contexts from which it emerged, by programmatic statements of its practitioners, and by the views of unsympathetic critics.

Postulates on repetitive travel form an axiomatic base for many classical theories of spatial structure. Rent and other mathematical models of urban land use involve assumptions about the tradeoffs associated with centrally oriented worktrips.[5] Central-place theory, market-area analysis and variants of location theory derive from corresponding assumptions of single-purpose, distance-minimizing consumer travel. During the 1960s dissatisfaction with the conceptually and geometrically rich structure of such theories focused on their stringent assumptions of the homogeneity and rationality of the actors involved. The desire to rewrite such theories in more palatable and positive, rather than normative, terms was an important incentive to the development of behavioral geography. This goal was acknowledged in broad programs such as that of Pred,[6] in more specific research designs,[7] and in explicit attempts to reformulate central-place theory in behavioral terms. It was a reasonable expectation that postulates on trip making would play as central a role in the new theory as they had in the old.

A second source of motivation for cognitive–behavioral study of travel behavior has been the search for disaggregate, psychological groundings for predictive gravity and intervening opportunity models. More or less explicitly, behavioral assumptions arise in attempts to find conditions on individual preferences, space discounting and utilities that are necessary and sufficient for gravity-like interactions,[8] and from the need for a microscopic interpretation of powerful aggregate ideas such as that of entropy.[9]

These theoretical and applied themes have provided external or contextual impetus for behavioral geographers to address repetitive travel. Virtually all programmatic statements within behavioral geography such as those found variously in Cox and Golledge,[10] Golledge and Rushton,[11] Downs and Stea[12] and Moore and Golledge[13] also imply that repetitive travel patterns can be fruitfully understood in cognitive–behavioral terms. A distinctive concern of this literature has been the description of cognitive maps; the activity patterns of repetitive travel, particularly path structures, are known to be a major organizing principle in cognitive imaging.[14]

The accomplishments and the overall advisability of the cognitive–behavioral paradigm have been questioned from many perspectives. Bunting and Guelke[15] maintain that the behavioral approach has failed to provide original and incisive accounts of overt behaviors of any kind and has, instead, been preoccupied with inaccessible, hypothetical constructs. Mental maps have been viewed as unproductive[16] and intrinsically flawed.[17] Behavioral geography has been described as an 'intellectual backwater,'[18] that is being 'vacated by geographers who evidently have decided the approach [is] not productive.'[19] The alleged failings of the behavioral paradigm have been traced to its quantitative–positivistic pedigree, to its failure to focus on subjective acts and meaning, to conservative adoption of questions posed in nonbehavioral geography, and to its purported failure to achieve either verification or replication.[20] Within the field, crosscurrents of criticism abound. For example, Louviere suggests that geographers have inadequately understood the bases of mathematical models borrowed from psychology,[21] while Michelson, although he supports the goals of behavioral geography, criticizes a tendency to develop labored mathematical justification of the obvious.[22] An implication common to criticisms as diverse as those of Bunting and Guelke, Cullen, and Dacey is that cognitive–behavioral approaches will or could be vindicated in geography by providing acceptable and distinctive accounts of overt travel behavior. (These authors differ profoundly in their definition of 'acceptable.')

This paper is an attempt to assess the impact of the cognitive–behavioral paradigm on geographic study of repetitive travel. This impact has been substantial, if it is measured by the pervasiveness of psychological, attitudinal, cognitive and decision terminology. It has been argued elsewhere that most behavioral studies of travel conform to a *spatial choice paradigm* that is inconsistent (or at least hard to reconcile) with the ideal program of behavioral geography.[23] The following sections review accomplishments in behavioral studies of travel and discuss refinements and criticisms of the paradigm approach. Finally, the standard paradigm is appraised in the broader context of the goals that were originally formulated for cognitive–behavioral geography.

Cognitive studies of destination choice

Choice is the dominant explanatory notion in most behavioral studies of travel. For example, Rushton's definitions of 'spatial behavior' and 'behavior in space' both explicitly involve choice.[24] A consequence of this pervasive focus on choice is a disproportionate emphasis in geographic studies on discretionary travel like shopping and recreational trips, rather than on travel that is space bound, such as the work-trip and most social travel. The work-trip is numerically the most important trip type for many adults. A concern for work travel was evident in early behavioral work in geography.[25] Workplaces, together with other fixed points such as schools and relatives' homes, are acknowledged to be primary anchors in overall cognitive images of urban space.[26] However, subsequent cognitive work has tended to neglect work travel in favor of discretionary trips. In contrast, the work-trip has been a dominant concern in disaggregate travel-demand modelling, though primarily from the viewpoint of mode choice – another topic that has been relatively neglected by geographers as a subject in its own right, although car ownership and other modal variables are frequently used as predictors of destination choice. A recent comprehensive review of studies of the journey to work contains no direct reference to cognitive–behavioral geography, though it does cite some literature on the perception of travel time and mode choice.[27] In effect, most cognitive studies of travel concentrate on shopping travel, with a subsidiary interest in recreation travel.[28]

This orientation has been termed elsewhere the spatial-choice paradigm.[29] Its main features are: *a priori* segmentation of trips from individual behavior; separation of point destinations from their objective context in real space and from their associative context in cognitive schemata; and a choice-based model of explanation that, at best, drastically simplifies the account of behavior provided by cognitive psychology. In constructing cognitive explanations according to the choice paradigm, geographers generally seek cognitive, attitudinal or other psychological transformations of traditional predictors of travel behavior, including attributes of sites, distance and locational variables, and relevant individual characteristics.

Cognitive transformation of site attributes

The two main site-specific predictors of destination choice in non-behavioral literature are size (measured by floorspace or number of offerings)[30] and price (e.g. delivered price, the principal predictor in central-place theory). Behavioral research has indicated that perceptions of sites are complex and multifaceted, and that neither size nor price is uniquely salient in preferences.

Two early and widely cited studies of the perceived attributes of retail destinations were those of Downs[31] and Burnett.[32] Downs analyzed attributes of a Bristol shopping center such as price, variety, service quality, and various contextual attributes, using the semantic differential and factor analysis. Burnett used scaling methods to compare clothing shopping in two groups with different information levels. The attributes studied included price, quality, variety, and a number of convenience factors including ease of parking and of exchanging goods. Cadwallader, in a study of supermarkets, addressed factors such as range of goods, speed of checkout and price.[33] Schuler employed a conjoint scaling model to data about grocery shopping.[34] Ratings were elicited on nineteen attributes divided into three groups: commodity traits such as quality and price, measures related to the stores themselves, and location measures including proximity to other specified activities. The most important variables were processed further, in constructing a weighted utility scale. These included the site specific factors of quality, price, and speed of service and nearness of parking. Lieber,[35] in an unusual study of choice of recreational (bowling) establishments used as site-specific variables the availability of refreshments and two measures of the availability of games. Patricios studied shopping for convenience goods including foodstuffs, hardware and personal and medical items in a South African setting.[36] Thirty-three attributes were specified, in three cognitive domains pertaining to sites, the shopping centers that contained them, and movement imagery. Ratings and rankings on the relative importance of attributes were collected. For example, cleanliness of premises, helpful personnel, quality and range of products were among ten scales considered 'extremely important.' Specific findings were provided on the perceived importance of the attributes for different types of convenience shopping.

In studies such as these a range of potential site attributes is specified *a priori*, although data are frequently compressed – for example, by factor analysis (Downs, Patricios), or by selection of a subset of the more important attributes (Schuler). Another strategy is to use methods that allow respondents to express what they consider to be important attributes, instead of merely rating prespecified dimensions. Explicit generation of attributes is

permitted by repertory grid analysis,[37] while Spencer used methodology that defines attributes in a more indirect way.[38] Spencer
studied grocery and produce shopping in England, using the multidimensional scaling techinique P R E F M A P. A convenience
dimension emerged that combined distance from the home and
variety of offerings. Another dimension related price to freshness,
an attribute peculiar to the commodities in question.

The studies outlined here, and many others constituted in the
same way, indicate the cognitive complexity of destinations in
discretionary travel. Again and again dimensions that are usually
thought of as clearly defined, such as quality or convenience, have
been shown to be multifaceted. Nonprice attributes such as
cleanliness, service quality and variety of offerings often exhibit the
strongest correlation with expressed preferences and revealed
behavior. These findings agree with results in marketing and travel
demand modeling. For example, Koppelman and Hauser report
that for nongrocery shopping trips the five basic aspects of
attractiveness are variety, quality, satisfaction, value and parking.[39]
Quality is the most important of these and store prestige seems to be
one of the most important components of quality.

Two principal difficulties arise in attempting to synthesize these
findings into conclusions on retail choice in general. Firstly,
attributes are often incomparable across different types of goods
and services. 'Quality' may evoke ideas of freshness and cleanliness
for grocery stores, and quite different aspects in, say, clothing
stores. This problem is compounded by an emphasis in geographic
studies on grocery and food supermarket trips. Many kinds of retail
functions have apparently not been studied at all, while findings on
drugs, clothing, hardware and the like are far too few to permit
useful generalization. One fruitful strategy is to attempt to
construct dimensions of attractiveness for stores in general, with
specialized components appropriate to particular functions (e.g.
the 'generalized attribute' approach of McCarthy).[40]

A second fundamental problem is that of disentangling store
attributes from attributes of the shopping centers or retail clusters in
which they are found.[41] Patricios indicates that in many cases people
form opinions of particular stores. On the other hand, contextual
characteristics such as parking and design are often shared by many
adjacent stores, and situational effects certainly arise in some cases.
This problem is compounded by the prevalence of multipurpose

trips for some types of goods and services. In this case, attributes of one store could conceivably be the salient reasons for a visit to an adjacent store of another type.

There is little consensus on technique in geographers' work on retail attractiveness. It is hard to find two studies with an identical methodology. Numerous attitude-rating, scaling and processing techniques have been used. There is some agreement that, among verbally orientated methods, repertory-grid/personal-construct theory is preferable to semantic-differential and Likert methods.[42] Others contend that nonverbal techniques such as variants of multi-dimensional scaling impose less *a priori* structure on responses.[43] Wohlwill expresses several other criticisms of personal construct theory as applied by geographers and environmental psychologists.[44]

Cognitive transformations of distance

Study of distance perception in its own right has been an important theme in behavioral geography.[45] Psychophysical ideas have been applied, including descriptions of perceived distance as a power or logarithmic function of real distance, usually in urban settings. Associated techniques such as ratio, direct-mileage and direct-magnitude estimation have been evaluated.[46] Systematic variations in such judgements have been traced to socioeconomic status, familiarity, location with respect to the city center, length of residence and other variables. Empirical testing of these findings usually involves comparing real distance or travel time with that predicted by a psychophysical model. These results stand somewhat apart from attitudinal approaches to repetitive travel. Only rarely are distance-perception studies phrased specifically in terms of destination choice. Conversely, studies of overt retail choice often use cognitive distance as a predictor, but rarely exploit methods advocated by Lowrey, Briggs, Cadwallader and others.

Despite the distinctness of psychophysical studies of distance and studies of attitudes toward destinations, cognitive transformations of distance are uniformly implicated as important variables in destination choice. For example, Schuler indicates that distance is among a handful of significant attributes in predicting supermarket preferences.[47] Pacione reports that incorporation of perceived distance increases the predictive power of a conventional gravity model.[48] Patricios indicates that distance and perceived access may be multidimensional variables,[49] while the findings of Spencer (outlined above) provide evidence that accessibility may be confounded with specific site attributes in a single cognitive dimension.[50] Overall, these results show the complexity of distance perception. They also indicate that time and measurable distance seem to be much less important than perceived comfort and convenience. Mode-choice literature indicates, for example, that time spent waiting is perceived as more costly than time in motion.[51]

Individual characteristics relevant to destination choice

In conventional and cognitive models of destination choice, controls on individual characteristics are almost always imposed. Logically, this may be done in two ways. The most straightforward is to classify respondents or households by variables such as socio-economic characteristics, car ownership, length of residence and locale, and to assume homogeneity of other characteristics within groups. Ideally, this process eliminates variation on the specified variables within groups and gives specific insights into the differences between them. This strategy is the one commonly adopted in geographic work. In extreme cases a desire to homogenize the sample very severely constrains its size. For example, in studying women's clothing purchases, Burnett imposed extremely stringent constraints on the location and characteristics of respondents.[52] The generality of conclusions from such studies is small, for obvious and unavoidable reasons.

The second strategy in aggregation is to incorporate uncontrolled variation into the mathematical structure of the choice models. Random utility models (and the logit model in particular) provide a means to accomplish this.[53] Pipkin indicates several different aggregation strategies of constant and random utility models which embody quite different assumptions of the homogeneity or heterogeneity of respondents.[54] If predictors such as attitudes to site characteristics are interpreted in cognitive terms, it is hard to conceive of any stratification that could eliminate residual variation. The relative merits of stratification and localization as opposed to an explicit model of variation are only rarely discussed in geographic work. Samples are typically made as homogeneous as possible by detailed stratification according to relevant individual characteristics and location with respect to potential destinations.

Socioeconomic differences in access and activity patterns have been well documented. Groups with higher status, higher income and higher mobility levels possess more wide-ranging activity patterns.[55] Distinctively cognitive analyses of these differences have emphasized the relationship between activity patterns and perceived action spaces, which are known to have a sectoral form in cities.[56] In several studies Potter has shown that the size and geometrical angle of information fields are positively associated with social class, measured by a classification of occupations.[57] In addition, the mean distance traveled and the mean number of shopping places visited were positively correlated with status. Lloyd and Jennings provide interesting evidence confirming an income difference in travel patterns, qualified by another variable, that of a possible ethnic bias.[58] It was found that higher-income consumers tend to travel farther for grocery shopping. A cognitive-rating model predicted lower-income consumers' behavior accurately, but did less well in predicting travel of the higher-income group. This disparity decreased when the ethnicity of clientele was taken into account.

These studies and many others stratify respondents by standard socioeconomic variables which do not possess an immediate behavioral interpretation. The logical next step in cognitive studies of travel would be to develop individual variables which possess direct psychological implications. An example of this approach is provided by Taylor.[59] Placing his work in the context of an extensive literature on environmental dispositions,[60] Taylor argues that certain personality dispositions may be effective in predicting spatial choice. Using information on clothing shopping in Vancouver he developed five 'orientations' – fashion, price, quality, convenience and status – which were incorporated into Likert scales. The dispositions were effective in differentiating two groups (students and shoppers). In predicting actual choices, the disposition scales were 'far superior' to locational measures but 'slightly inferior' to standard socioeconomic and demographic predictors. Overall, the strongest predictors of choice were fashion and status orientation, and household income.

In most geographic studies individual characteristics are constituted *a priori* using standard criteria of location and socioeconomic status, without raising the question of which individual variables are most effective in differentiating travel patterns. However, there is a growing interest in this problem of segmentation. It is a long-standing concern in marketing and consumer research. An early foray into this area by geographers was provided by Louviere *et al.*[61] Subsequent work in travel-demand modeling has emphasized the importance of attitudes, preferences, intentions and perceived constraints in segmentation.[62] In a study of choices for recreational travel, Stopher and Ergun suggested that variations in tastes, motivations and personalities could be usefully summarized by segmentation of location, family life cycle, and of variables associated with destination attractiveness.[63]

Two modes of explanation in the choice paradigm

Formal models of choice typically contain two components: an account of how predictive dimensions are combined (for example, in an intervening scale) and a model that stipulates the role of the scale values in an individual act of choice. These may be termed response and choice models, respectively.

A crucial dichotomy in the theory of choice distinguishes algebraic and probabilistic choice models.[64] Quite distinct mathematical structures are appropriate to these two models. The former involves binary preference relations, algebraic orderings and deterministic utility functions. The latter involves notions of stochastic transitivity and various utility constructs including weak, strong, strict and random utility models.

It has been argued elsewhere that geographers have not adequately addressed this distinction and that a probabilistic account may be more appropriate for description of destination choice.[65] An algebraic approach is implicit in many geographic accounts of destination choice,[66] although there seem to have been no studies at all that have attempted to compare empirically the appropriateness of these two perspectives on a specific individual act of choice. In travel-demand modeling the probabilistic paradigm is fully established,[67] and there is an emerging consensus that it may be appropriate in modeling repetitive travel.[68]

To date, geographers' work on destination has in effect focused on the response function. The preceding sections indicate that substantial progress has been made in identifying and measuring cognitive aspects of sites, of trip-links, and of relevant individual characteristics. The internal logic of the choice paradigm suggests that in future more emphasis will be placed on appropriate choice models. If probabilistic choice models are in fact most appropriate to understanding repetitive travel, future geographic work must evaluate structures such as constant and random utility models,[69] Luce's choice theory,[70] its alternatives[71] and its extension,[72] its limit properties[73] and its aggregation potential.[74] Geography is likely to benefit from the intensive debate and criticism of these models in travel-demand modeling, which has focused on the principal probabilistic models of travel choice: multinomial logit and probit methods.[75]

Spatial choice paradigm: problems and criticism

The preceding section suggests that the spatial choice paradigm has not yet been carried through to its logical conclusion, the formulation and empirical evaluation of mathematical models of choice at an explicitly individual level. This aspect of the paradigm has been carried farther in disaggregate travel demand modeling than in geography.[76] Planners, economists and engineers engaged in this work have studied repetitive travel in specifically psychological terms supported by levels of funding and sample sizes inaccessible to most academic geographers.[77] Citations in transportation research and planning literature suggest that the work of few North American geographers is known to these researchers. Studies of holistic cognitive images appear to have been acknowledged in transportation planning practice only quite recently.[78] The theoretical foundations and the mathematical techniques of disaggregate transportation planning have been drawn from new ideas in microeconomics and, particularly, from choice theories of psychology, rather than from geography.

There are evidently institutional differences and divergences in goals between behavioral geography and psychologically oriented travel modeling. However, studies of destination choice in geography possess very strong formal similarities to those in travel-demand studies. The choice paradigm in both fields has recently been subject to criticisms by geographers and transportation planners; appraisal of these criticisms is essential in evaluating the potential for cognitive–behavioral accounts of repetitive travel.

Two principal problem areas in destination-choice models are the divergence of reported and revealed preferences, and a growing appreciation of the space–time complexity of repetitive travel.

Revealed and reported travel preferences

Almost uniformly, predictive studies of destination choice report some divergence between predicted and observed behavior,[79] and also between elicited preference rankings and the rankings determined from inferred utility scales, even for the same respondents.[80] Interpretation of such results has always been problematical. The well-documented discrepancy between attitude and act is a perennial difficulty in attitudinal studies, particularly those that use complex questionnaire instruments, or which confront individuals with prespecified evaluation criteria (e.g. most applications of the semantic differential) rather than less obtrusive techniques that in principle allow individuals' evaluation criteria to emerge inductively (e.g. repertory grid). This divergence can be used to motivate an approach to behavior that attempts to infer intervening cognitive variables from revealed behavior exclusively.[81] In the context of probabilistic models a strict revealed-preference approach can be shown to constrain very severely the kinds of inferences available at an individual level, mainly because of the rarity with which households reveal certain kinds of destination choices for shoppers' or specialty goods.[82]

The most straightforward explanation of the divergence between act and attitude is simply that people are inconsistent in their behavior, or that current questionnaire techniques are unable to elicit their true attitudes. 'The fundamental question remains whether it is possible to obtain subjects who are prepared and able to reveal their actual thoughts to investigators.'[83] There does exist some evidence that people believe distance to be more significant a determinant of travel than it actually is,[84] and that price is perceived as more important in choice than it really is.[85]

Another possible explanation of the predictive failings of inferred attitude lies in specification errors of various types. Firstly, predictors which have been correctly identified may differ in their salience for different individuals, even in socioeconomically homogeneous and localized samples. Geographic studies typically assume comparability (constant weights) of predictors across individuals and sites. It is logically possible to handle individual- and site-specific predictors in such models as the multinomial logit.[86] Another possibility is to employ scaling techniques, such as PREFMAP, that assume comparable underlying dimensions, but which accommodate individual variations.

Secondly, even if predictor variables are correctly specified, the combination rule that maps them to a (conventionally one-dimensional) utility scale may be incorrectly specified. Usually simple multiplicative or additive separability is imposed. Louviere and Lieber indicate the difficulties of inferring combination rules from real data, and emphasize the different strengths and weaknesses of techniques such as polynomial-conjoint and functional analyses.[87] Inferences on combination rules require data in which alternatives exhibit a wide range of values on each attribute. Studies which seek a combination rule with some degree of generality have recourse to hypothetical alternatives rather than actual destinations, as in Louviere's and Lieber's studies. The exigencies of real data frequently require a simple, usually additive, combination rule to be specified by the researcher. In such cases, the potential for specification error is inherent in the limitations of data, or even in urban structure itself, where destinations of a specified type may simply not exhibit sufficient independent variation in attributes to permit delicate refinement of the combination rule from revealed preferences alone. It is often necessary to make simplifying assumptions on separability (for example, separability of site and distance attributes in utility functions) when it may be empirically unwarranted.

Another quite different perspective on the divergence between act and attitude emphasizes the bias, amounting to social class bias, in imputing free choices where few exist. Evidence clearly shows that repetitive travel patterns vary significantly across income, status, ethnic and other groups, as well as for handicapped, elderly and other specifically disadvantaged people. Several authors, including Spencer,[88] suggest that attitudinal and preference scaling procedures frequently elicit individuals' true preferences, but that, because of constraints on income and mobility, they cannot be attained. He illustrates this assertion with evidence about English food-shopping patterns. The constraint interpretation of the disparity between preference and behavior is described by Sheppard as an inadvertent ideological bias in travel modeling, which has the effect of reinforcing existing social and urban structure.[89] In principle, constraints could be built into existing choice models by careful specification of attainable choice sets, and by appropriate pricing of distance and time, within budget constraints. In general, though, the constraint problem provides a compelling argument against a strict revealed-preference orientation, since desired but unattainable choices are never observed.

Study of the relationship between attitudes to sites and to modes and real travel behavior is extremely active in travel-demand literature.[90] Technical innovations in measuring the importance of attributes are typified by Johnson's work, and by that of Levin and Lerman in joint work with Louviere, who stress inferences on the functional form of the combination rule. Dobson *et al.* address the question of the development of attitude and behavior over time in the context of bus travel in Los Angeles. This temporal view suggests that observed discrepancies between act and attitude may represent a (perhaps temporary) lack of adjustment between two structures evolving at different rates. Another potentially important approach to resolving the attitude–behavior problem is proposed by Hensher and Louviere, who introduce a mediating concept: behavioral intention toward specific alternatives.[91]

These perspectives on the relationship between attitudes and behavior indicate clearly that an understanding of the cognitive transformation of site and location attributes, no matter how detailed, is insufficient in explaining repetitive travel. In cases where choice may be inferred, these findings reinforce the conclusion of a preceding section: knowledge of cognitive attributes needs to be supplemented with explicit choice models of the combination rules and other mediating transformations that link attitudes to behavior. And, in general, we need more information on the constraints that may render irrelevant choice models of any kind.

Criticisms of the choice paradigm from the structure of behavior

Far-reaching criticisms of the paradigm approach to destination choice have been expressed by geographers and transportation planners. Burnett, Ellerman, Hanson, Heggie, Heggie and Jones, Hensher, Van der Hoorn[92] and others have implied with various degrees of emphasis that in its simple form the choice paradigm is incapable of addressing the complexity of repetitive travel. Heggie asserts that, despite their psychological trappings, conventional choice models are not behavioral at all. These criticisms stem from evidence on the spatial structure of activity patterns, and on the behavioral context of travel, with special emphasis on temporal and other constraints that people contend with in everyday activities.

Distinctively spatial evidence on the inadequacy of the standard choice paradigm arises in the study of multipurpose travel. Such trip structures have been analyzed in their own right, mainly with emphasis on the strength of linkages between various types of establishment.[93] The fundamental problem of multipurpose travel from a decision-making perspective is that potential stops acquire complex, opportunistic utility by virtue of proximity to other sites. Hanson provides various insights into the modifications needed if choice models are to address this problem.[94] She provides evidence that visit frequencies and inferred preferences for sites within a function category do vary on single and multipurpose trips. Some functions actually possess stronger trip linkages to the workplace than to the home or to other establishments. Hanson indicates that traditional categories of data, and modes of collection, have tended to minimize the importance of such trips. Detailed longitudinal information that distinguishes stages of journeys is required to study the time and space aspects of multipurpose trips.

In most geographic studies of destination, choice sites are characterized by their location relative to respondents' homes. Schuler did elicit ratings on proximity to sites such as banks, drugstores and discount stores as a component in the attractiveness of grocery stores.[95] In this case none of these variables emerged as an important predictor of either preference or choice. Few geographers have attempted to spell out formal models of multipurpose trips along the lines of work in travel-demand modeling.[96]

A complete understanding of spatial patterns in multipurpose travel requires an appreciation of the temporal and behavioral context of such trips. The time dimension of repetitive travel has been neglected in most studies of repetitive travel, though Curry indicated the importance of household demand cycles and inventories of various goods as a determinant of multipurpose trips.[97] More recently, study of various aspects of time geography has generated interest in the temporal context of travel behavior. It has become clear, for example, that aggregate urban images depend on time. Tranter and Parkes used repertory grid analysis with twenty-five standard elements (including shopping, recreational, religious and other sites) and with ten standard attributes (such as liking, newness, safety and business) to exhibit dependency of image and evaluation in time, and to describe 'space–time drift' of elements.[98] Parkes and Thrift have emphasized the need for a unified treatment of space and time,[99] an approach explored in various works by Hägerstrand and his associates in which trips, including multi-purpose trips, are represented as space–time entities on coordinate diagrams.[100] A principal concern in space–time studies of travel is the constraint imposed on trips by the scheduling of household activities in free time and committed time (e.g. work or school). A discussion of this constraint-oriented perspective behavior is provided by Hensher.[101]

The transportation-planning literature on space–time and constraint-oriented models communicates a sense of a new beginning, of a third generation of planning models to supersede aggregative and 'disaggregate-behavioral' approaches. As argued above, the disaggregate-behavioral tradition represents a more mathematical equivalent of the spatial-choice paradigm in geography. This new beginning raises two concerns shared by planner-analysts and by a handful of geographers: firstly, to circumscribe a legitimate domain of application for highly developed existing models, notably the multinomial logit model; and, secondly, to develop appropriate new data and models.

Heggie and Jones attempt to indicate appropriate domains of applicability for various models,[102] arguing from a review of empirical evidence on households' responses to planning policies which require various degrees of mandated or voluntary reorganization of activities such as car restraint, traffic restraint, flexible working hours and rescheduled school hours. Potential responses to such policies include reallocation of time, and reorganization of household travel patterns and of the division of labor in the household. Considering possible spatial, temporal and interpersonal dependencies, Heggie and Jones specify four model domains. The domain of *independence* is that assumed in traditional spatial and modal choice models in which destination choices and other decisions are largely unbound by time, space, or the needs of others. This domain may be most appropriate to the activities of well-to-do, childless adults. Other model domains address *personal* linkages, *spatio-*

temporal linkages and full *interdependence*, the most common and realistic case. In each domain except the first, new data and conceptualizations are required.

The new orientation in travel studies entails a new empiricism, in which travel patterns are described longitudinally with reference to constraints, new survey techniques, and an appraisal of the formal structure of predictive models. New forms of data description are exemplified in space–time diagrams, which Burnett and Hanson study from an inductive point of view, attempting by flow analysis and various taxonomic ideas to identify significant patterns in path structures.[103] Typical of new survey methods is the Household Activity Travel Simulator.[104] This method, applied with a detailed interview, probes the structure of household activities with a display board in which activities are blocked and color coded. When used in conjunction with travel diaries and other data this method reveals activity patterns in great detail. Unlike standard methods, it can be used to elicit likely responses to proposed planning policies. Burnett and Ellerman provide a detailed prospectus of new and old methods including large-scale sampling, travel diaries, a household interaction game, specific studies of choice sets and decision procedures, and participant observation.[105] Some of these methods are well tried in transportation planning, though they have engaged the attention of few geographers concerned with repetitive travel.

There is less consensus on the mathematical or formal structure of suitable predictive models, or even on the need for mathematization. Burnett and Hanson outline, in highly abstract formalism, a probabilistic model involving a behavior complex, choice set constitution and conditional site selection.[106] Hensher proposes an analogous formalism involving land-use configurations, 'routine influences' and mode choice.[107] On the other hand, Heggie[108] and Cullen[109] both imply that 'models of behavior may have to become more human and less mathematical.'[110]

Repetitive travel and the goals of cognitive behavioral geography

The preceding sections outline two quite different criticisms of the standard geographical paradigm of destination choice. Firstly, within the frame of reference of the choice paradigm, insufficient attention has been devoted to choice processes at a specifically individual level. Many studies are conducted in a way that does not facilitate an empirical decision between probabilistic and algebraic accounts of specific acts of choice. It seems likely that a probabilistic model is appropriate. In this case constructs such as binary preference orderings, utility scales and algebraic conjoint scales need to be superseded by their probabilistic analogues: weak, strong, strict and random utility models, and stochastic ideas of ordering and transitivity; and probabilistic choice models of mathematical psychology. This evolution has in fact occurred in behavioral travel-demand modeling, where the multinomial logit is a standard choice model. On the other hand, critics have maintained that the choice paradigm is inadequate to capture the relationship between attitude and behavior; that to impute choice in repetitive travel may be incompatible with a concern for social justice; and, above all, that the spatiotemporal and behavioral contexts of travel have been over-simplified.

Only a few North American geographers have been involved in re-evaluation of probabilistic spatial choice models, and in developing constraint- and time-oriented account of travel. Core work in cognitive geography, on mental images of urban space for example, has played a relatively small role in informing this criticism or in suggesting new conceptual frameworks. Instead, refinements in travel methodology have stemmed from empirical descriptions of the structure of travel, from consideration of household decision making, and from criticism of rationalistic choice models.

Within geography, studies of travel, particularly discretionary trips, commonly employ behavioral and psychological terminology, but the structure of explanation in the choice paradigm has remained intact. Cognitive ideas have led only to the conservative strategy of seeking psychological transformations of traditional predictors that figure as independent variables, while single trips or elicited preference rankings provide dependent variables. There is a fundamental discontinuity between this mode of explanation and the goals originally formulated for cognitive behavioral geography, which may be inferred rather specifically from statements in Rushton, Cox and Golledge, Moore and Golledge, Downs and Stea and other works cited above. The three fundamentals of this program were:

1 An account of behavior and mental activity consistent with that of the *cognitive* school of psychology;

2 Primacy of inferred *cognitive constructs* (e.g. mental maps) as both products and determinants of behavior; and
3 A desire to develop *postulates* on behavior that are general, cognitive and in some sense spatial.[111]

Cognitive explanation

The cognitive paradigm of psychology construes behavior as a continuum of acts arising jointly from external stimuli and internal schemata, and propelled by goals, plans, values, drives and motivations. Behavior is mediated in conscious thought by attention, and at deeper levels of short- and long-term memory where inaccessible processes occur, including, certainly, the decay of information and, probably, the formation of associations and processing between semantic and other modalities.[112]

A distinctive feature of the cognitive paradigm, which is inadequately recognized by geographers, is the structuring of behavior not by observable, spatial or topographical properties, but by inferred motives and intents. Another feature of this approach is wholehearted acceptance of mentalistic ideas, particularly imagery, eschewed by other paradigms of psychology.[113] The causal or explanatory power of external variables arises not directly from their observed correlation with acts, but from demonstration that they are implicated in mental activity either directly, in consciousness, or more remotely in the formation of habits and of plans. It has been argued at length elsewhere that it is hard to reconcile geographical work on travel with the cognitive paradigm.[114]

Geographers' central explanatory idea of spatial choice is hard to define from a cognitive perspective. Choice, or even the illusion of free will and control, is known to be related to well being and to the quality of performance in various tasks.[115] Yet our mental lives contain all kinds of decision-like events as we monitor our position and progress toward mundane and overall goals, stemming proximately from plans, and more remotely from acquired drives, value systems and habits. Situations that require deliberation, which elicit stress and ambiguity, are quite rare, and cannot be related in any simple way to the places we visit in everyday travel. It is clear that the notion of spatial choice as developed by Rushton[116] and in Golledge and Rushton[117] needs refinement. Steiner provides a taxonomy of choices as discriminative, evaluative and autonomous, that may provide clues to the refinements required.[118] It is also plain that the data used in most studies of destination choice[119] are quite inadequate either to classify the contexts of travel behavior by ambiguity, stress or 'freedom,' or to demonstrate the salience of schemata and stimuli to acts in specifically cognitive terms. A simple dichotomy of 'habitual' or routinized and 'genuine' or innovative choices is also inadequate to understand the behavioral context of repetitive travel, since even habitual trips are likely to be linked to stress-eliciting choices concerning timing, car availability, income constraints and the like. Thus in the specific context of decision making, and in the more general task of showing relationships between ongoing mental activity and relevant behaviors, geography has failed to satisfy the requirements of cognitive explanation.

Internal schemata and repetitive travel

The most distinctive feature of core work in behavioral geography is emphasis on inferred cognitive schemata ('mental maps'). The principal methodological trend in the first decade of this paradigm has been refinement in the means of eliciting and representing such constructs (for example, by multidimensional scaling).[120] These constructs are not well integrated into behavioral studies of destination choice or other repetitive travel, nor have studies of specific overt travel patterns been prominent in the literature on mental maps. The discontinuity between these literatures has at least three aspects, concerning the role of values, the geometric structure of images, and their explanatory or causal role. Firstly, mental maps are often inferred in designative or strictly cognitive terms. Attempts to identify preferences associated with general urban imagery by rating techniques or repertory grid have shown that these affective dimensions are very varied.[121] By contrast, studies of destination choice have strongly emphasized affective aspects of image, usually in terms of attitudes toward price, convenience, travel time and similar standardized dimensions. Secondly, mental maps are usually represented in holistic form, as sketch maps or as MDS configurations. In destination choice studies, relevant predictors are associated with point-like sites characterized only by a few geometric measures such as distance to home, and perhaps by some contextual measures related to the setting of sites in retail clusters or shopping centers. The spatial or geometric richness of cognitive representations of space is not usually built into models of destination choice. Thirdly, in studies of travel, attitudinal and other 'cognitive' variables are usually applied as independent or explanatory variables, in relation to dependent choices or preference rankings. In studies that have focused on cognitive maps, the contents of images are usually treated as dependent variables, in relation to variables such as socioeconomic status, length of residence, mobility characteristics and activity patterns.

The relationship between internal schemata and revealed behavior is extremely complex, considered from a cognitive perspective. To begin with, the internal structure of images is by no means as clearcut as implied in early literature on mental maps. The idea of *imagery*, often invoked more or less formally by cognitive geographers, has reemerged as an important theme in psychology and psychological work indicates some of the difficulties in using image concepts to explain behavior. Paivio identifies two strands of research concerning images, theoretical and empirical.[122] Theoretically, images are treated as mental events with visual or pictorial properties, which are implicated in 'storage and symbolic manipulation of information concerning spatially organized objects and events,'[123] and which are parallel-processed in a relatively fast and crude fashion. Verbal or semantic coding, on the other hand, is associated with auditory and motor systems and is 'specialized for sequential processing.' Empirically, according to Paivio, image studies have focused on the characteristics of stimuli including their clarity and concreteness (viz. Lynch's 'legibility'); the way in which imagery facilitates problem solving, learning and performance (e.g. the visualizing tricks in mnemonic systems); and the delineation of individual differences in capacity and preference for visual and verbal thought. Kaplan,[124] Moore and Golledge[125] and others have alluded to the importance of these distinctions in geographic-image studies.

Attempts to explain repetitive travel with reference to cognitive images must inescapably address internal processes including the formation and decay of associative networks; the processing that occurs between deep levels and the surface images that arise in thought and decision making; and the interplay of semantic and image modes. Geographers have shown little concern for modeling internal processes along the lines of work by Beck and Wood,[126] who make specific suggestions on cognitive transformations such as representation, coding, recording, schema formation and retrieval. One potentially serious hurdle in this endeavor is the profound difference in the processes involved in responding to questionnaires and interviews and those arising in real repetitive travel. The difference is akin to that between recall and recognition, which is a most significant one in the psychology of memory.[127] Real travel certainly involves much visual, recognition memory. Questionnaire responses about travel patterns, shopping centers and the like require recall, and evokes thought processes such as distance estimation and rating judgements that are quite alien to most people's everyday thought. In most interview techniques, including those of semantic differential and repertory grid, a final verbal product is elicited, which is, of course, untrue to real travel. A host of distinctively psychological ideas such as association, latency, interference, concreteness and, not least, meaning, could be implicated in rigorous study of these processes. Cognitive geographers are well informed of these problems, but no existing studies of repetitive travel address internal processes with any rigor, or in a fashion that could identify the causal role of specific schemata with any certainty. In fact, there is only one case in which clear causal priority of image to behavior might be inferred: the case of innovative behavior, perhaps involving search for a rarely needed item or service. In behavior that is routine (in the sense of spatial regularity), such as the work-trip, it is more likely that small incremental changes in images are results and not predictors of trips. Evidently, a simple dichotomy of dependent and independent variables cannot do justice to the continuing transaction between behavior and image, though such a structure is clearly implied in the statistical structure of destination choice models.

The search for behavioral postulates

Programmatic statements for cognitive geography, such as those of Harvey[128] and Cox and Golledge,[129] identified 'explicit assumptions about human behavior' as a necessary foundation for explanation in behavioral geography.[130] At least four requirements seem to have been set for such postulates. They should, firstly, be in some sense cognitive rather than normative and rationalistic. Secondly, they should be general and not bound to specific spatial structures (in the sense of being 'independent of the spatial system in which they have been observed'.[131] Thirdly, they should in some sense be spatial or geometric. Finally, they should not be idiosyncratic, but should be applicable to many individuals. Thus postulates are required to be cognitive, abstract, spatial and general, in addition to satisfying implicit requirements of simplicity, theoretical fruitfulness and testability.

The need for an axiomatic base has not been emphasized in recent work on the structure, formation and processing of cognitive maps. Golledge provides one of the few examples of an attempt to isolate the required assumptions.[132] He emphasizes the essentially constructive role of the mind in its transaction with reality, making explicit some of the primitives that are logically necessary for such constructions to arise. The resulting perspective is one of a dialectical process between individual and environment, in which the central questions concern the kind of transformations made by the mind.

Most recent work, and most innovations in technique, do not seek an economical set of assumptions applicable to various thoughts and actions. Instead, individual differences are emphasized. Methods such as personal-construct theory are designed precisely to allow scope for such differences, and to avoid *a priori* imposition of the elements of images, or of attitudes toward them, of the kind implicit in, say, semantic differentials with prespecified scales. There seems to be some reason to doubt that postulates satisfying the four requirements above can arise inductively in strictly cognitive studies of spatial images and behavior. A deductive, axiomatic orientation does not seem common in cognitive psychology. Deductive systems that have been developed, such as those concerning idealized information processors,[133] involve primitives and inferences about internal schemata and the retrieval, processing and capacity characteristics of mental activities including perception and memory. The cognitive perspective emphasizes the inseparability of urban images and 'geographic' behavior from all other images (in associative context) and from all other behavior (behavioral continuity). There is to date no evidence that a general axiomatic system for travel behaviors can be developed that is strictly cognitive and distinctively geographic.

The third requirement mentioned above, that geographic behaviors should be constituted in a spatial or geometric way, is particularly problematic. Some psychological paradigms emphasize decompositions of behavior by observable structure that admits, in principle, of geometric description. B. F. Skinner's definition of 'operant' seems to be essentially topographical ('The properties which define a response are observable data').[134] This is not true of the cognitive paradigm, in which behavior is constituted as a continuum of motivated acts embedded, perhaps hierarchically, in various goals that may be pursued simultaneously, or may be continually interrupted by higher-priority goals. Even sophisticated space–time topographies of single and multipurpose trips, such as those discussed by Burnett and Hanson,[135] are intrinsically non-cognitive in this respect. It is of course necessary to distinguish the external representability of cognitive constructs in geometric or map-like form, from the far stronger and debatable assertion that internal associative networks are in some sense geometric. As Olsson, Svart and others have implied, constitution of overt behaviors by their observable structure may simply not permit the required 'causal' inferences.[136] Not only is it true that 'behavioral postulates cannot be deduced from behavioral patterns,'[137] but *a priori* spatial specification of behavior may be irreconcilable with genuinely cognitive explanation.

Deductive or axiomatic reasoning is rare in geographical studies of repetitive travel. The only postulates commonly mentioned in empirical work are those derived from a conservative 'psychological' reformulation of microeconomics (for example, Cadwallader invokes the idea of bounded rationality).[138] There is little evidence that choice and preference ideas will or should be central to cognitive explanation, as suggested above. Ideological and other failings of the choice-oriented cognitive reconstruction of rationality have been indicated by critics of the choice paradigm.[139] In extreme cases the inferred postulates may be vacuous in that they do not constrain data in any conceivable way. This would be so in a study which inferred preferences consistent with an observed pattern of responses and failed to predict any new cases. As Sheppard indicates, the imputed egotism may be empirically false or, when behavior is tightly constrained, irrelevant. At best, it is clear that empirical studies of destination choice subsist on postulates that clearly bear traits of their normative microeconomic prototypes, and which are not founded on specifically cognitive findings.

Smith has made one of the few attempts to develop a new

axiomatic base for the study of repetitive travel.[140] He introduces three postulates: an *independence* axiom analogous to Luce's choice axiom, which requires that addition or deletion of elements from the choice set have no effect on the relative choice probabilities of other elements;[141] a *separability* axiom, which requires independence of site and locational effects; and an *accessibility* axiom, which introduces geometric content by requiring that a diminution in real or perceived distance shall not decrease the corresponding choice probability, *ceteris paribus*. These assumptions are shown to imply restrictions, for example on the functional form of distance deterrence. This approach is attractive for its deductive structure, its specifically spatial content, its appeal to psychology rather than to microeconomics, because its account of individual acts of choice is probabilistic, and because remote and interesting consequences have been deduced from few, simple and appealing postulates. However, Smith's approach remains choice oriented, the substantive acceptability of Luce's axiom is debated in psychology, and the approach is in no sense cognitive.

The most natural way to fuse the choice and the cognitive-map schools of behavioral geography may be to concentrate on the cognitive processes of decision making rather than on the structure of schemata as such. An example of a relatively conservative effort along these lines is provided by Hanson, who retains a traditional choice perspective and attempts to show the modifications required by findings on multipurpose travel.[142] As indicated above, though, a cognitive account requires a thorough refinement of choice concepts and the development of data capable of exploring the true subjective context of decisions with reference to deep and surface processes of the mind. In developing such a system of assumptions on decision making, it seems necessary to draw a clear distinction between cognitive and behavioral postulates.

Cognitive postulates must address the conscious processes of spatial problem solving, the correlates in memory of features of urban landscapes such as concreteness, associations and meaning generally,[143] the significance of semantic coding and transfer,[144] and above all the kinds of *thought* surrounding real travel behavior. As Tuan pointed out in a most incisive paper, the cognitive program entails mentalism to the extent of recognizing the primacy and meaning of 'objects' and 'events' in conscious thought.[145] Some authors, including Cullen, are sceptical of the utility of the traditional positivist tools of behavioral geography in this endeavor.[146] My reading of cognitive psychology on the topics of imagery and meaning, admittedly that of an outsider, suggests that very considerable progress may be made using objective reports and measures, replication and the usual procedures of statistical inference. It is clear, though, that the combination rules, weighted utility functions and linear preference orderings of the spatial choice paradigm are indefensible as mirrors of 'real' (experienced) processes in conscious thought or habit formation.

Behavioral postulates, on the other hand, are likely to involve primitives derived from *a priori* spatial constitution of behavior and urban structure, informed by findings on the space–time complexity of such action; and they might well involve powerful hypothetical constructs such as utility scales and orderings, and behavioral assumptions such as those of Smith, or others in the spectrum of behavioral hypotheses outlined by Cesario and Smith.[147] Formal, probabilistic choice theories such as those of Luce, Tversky and others provide models of the required amalgam of inductive and strictly axiomatic reasoning, of a kind that is extremely rare in geographic studies of travel. The problem, phrased in these terms, is to connect the cognitive and the behavioral in geography.

Conclusion

In traditional theories of urban and economic geography, assumptions on repetitive travel yielded unified accounts of the behavior of consumers and producers, and of the mutual relationship between individual behavior and aggregate spatial structures. No cognitive–behavioral equivalents of these theories have been produced. For example, very few cognitive studies have attempted to connect the behavior of consumers and entrepreneurs.[148] In assessing the actual impact of cognitive ideas on the analysis of repetitive travel it is necessary to distinguish two strands of work in behavioral geography; the paradigms of spatial choice and of cognitive maps. The orientations of these approaches are quite different and there is little sign of an impending synthesis.

Choice approaches were developed by many geographers, but notably by Rushton and his associates. Such studies accept the structure of behavior (trips), and the dichotomy of dependent (choice) and independent (attribute) variables in previous geographic work on travel. Explanation hinges on inferred choices, subjective utilities and preferences. This approach evidently represents a line of least resistance in modifying existing theories, with which it is highly compatible.[149] Continuing technical innovations in direct and conjoint scaling have produced rapidly growing, though somewhat incoherent, knowledge of the cognitive dimensions of destinations, and have established statistical relationships between attitudes to sites and trip frequencies. The choice paradigm has had a pervasive effect on studies of discretionary travel. It is acknowledged in a majority of papers on shopping and recreational trips. The preceding discussion suggests that the logical next step within this paradigm is to move from the current preoccupation with attribute dimensions to consider probabilistic and mathematically explicit models of individual acts of choice. This will involve appraisal, by more geographers, of structures such as the multinomial logit model that have been extensively used, and criticized, in a decade of travel-demand modeling. Various criticisms of the spatial-choice paradigm have been outlined. Its constitution of behavior does not do justice to the complexity of real travel, and its constitution of choice is quite inadequate as a replica of real cognitive processes.

The second strand of work in cognitive geography, concerning mental constructs, was developed by Cox, Downs, Golledge, Gould, Rushton and many others, and builds on ideas of Kenneth Boulding, Kevin Lynch and Anselm Strauss. It represents a quite radical departure from traditional geography. As yet it has had little impact on studies of overt travel, although travel patterns have provided much insight on the form of mental maps. It is argued, for the reasons given above, that the discrepancy between cognitive explanation using schemata and studies of spatial choice is far greater than commonly realised. Specifically, there is no sign of the distinctively cognitive, axiomatic base, hoped for in the formative period of cognitive geography. Synthesis of the choice and cognitive perspectives seems essential before the extensive materials on cognitive schemata can be brought to bear on overt travel in significantly new ways. Such a synthesis is likely to include a thorough refinement of theory and data concerning choice, more explicit study of mental processes and thought, and a clear distinction between behavioral and cognitive postulates.

Notes

 1 A. Wilson (1972) 'Some recent developments in microeconomic
    approaches to modeling household behavior, with special reference
    to spatiotemporal organization,' in A. Wilson (ed.) *Papers in Urban
    and Regional Analysis*, London, Pion; A. Wilson (1973) 'Further
    developments of entropy-maximizing transport models,' *Transport-
    ation Planning and Technology*, 1, 183–93; A. Wilson (1974) *Urban
    and Regional Models in Geography and Planning*, London, Wiley.
 2 D. Ley and M. Samuels (1978) *Humanistic Geography: Prospects
    and Problems*, Chicago, Maaroufa.
 3 D. Harvey (1973) *Social Justice and the City*, London, Arnold;
    L. King (1976) 'Alternatives to a positive economic geography,'
    Annals of the Association of American Geographers, 66, 293–308.
 4 For example, R. Davies (1969) 'Effects of consumer income differ-
    ences on shopping movement behavior,' *Tijdschrift voor Econo-
    mische en Sociale Geografie*, 60, 111–21; R. Potter (1977) 'Spatial
    patterns of consumer behavior and perception in relation to the social
    class variable,' *Area*, 9, 153–6; R. Potter (1977) 'The nature of
    consumer usage fields in an urban environment: theoretical and
    empirical perspectives,' *Tijdschrift voor Economische en Sociale
    Geografie*, 68, 168–76.
 5 For example, W. Alonso (1964) *Location and Land Use*, Cambridge,
    Mass., MIT Press.
 6 A. Pred (1967) *Behavior and Location, I*, Lund University, Studies in
    Geography, B (27).
 7 F. Horton and D. Reynolds (1971) 'Effects of urban spatial structure
    on individual behavior,' *Economic Geography*, 47, 36–49.
 8 J. Niedercorn and B. Bechdolt (1969) 'An economic derivation of the
    "gravity law" of spatial interaction,' *Journal of Regional Science*, 9,
    273–82; W. Isard (1975) 'A simple rationale for gravity model type
    behavior,' *Papers of the Regional Science Association*, 35, 25–30; T.
    Smith (1975) 'An axiomatic theory of spatial discounting behavior,'
    Papers of the Regional Science Associaton, 35, 31–49; T. Smith (1975)
    'A choice theory of spatial interaction,' *Regional Science and Urban
    Economics*, 5, 1–40.
 9 Wilson, 'Further developments of entropy-maximizing transport
    models' (see note 1 above); M. Webber (1977) 'Pedagogy again:
    what is entropy?' *Annals of the Association of American Geogra-
    phers*, 67, 254–66.
10 K. Cox and R. Golledge (1969) *Behavioral Problems in Geography:
    A Symposium*, Evanston, Ill., Northwestern University, Department
    of Geography, Studies in Geography 17.
11 R. Golledge and G. Rushton (eds) (1976) *Spatial Choice and Spatial
    Behavior*, Columbus, Ohio State University Press.
12 R. Downs and D. Stea (eds) (1973) *Image and Environment*,
    Chicago, Aldine; R. Downs and D. Stea (1977) *Maps in Minds*, New
    York, Harper & Row.
13 G. Moore and R. Golledge (1976) 'Environmental knowing:

concepts and theories,' in G. Moore and R. Golledge (eds) *Environmental Knowing*, Stroudsburg, Pa., Dowden, Hutchinson & Ross, 3–24.

14 See, for example, K. Lynch (1960) *The Image of the City*, Cambridge, Mass., MIT Press; and A. Devlin, 'The "small town" cognitive map: adjusting to a new environment,' in Moore and Golledge, *Environmental Knowing* (see note 13 above).

15 T. Bunting and L. Guelke (1979) 'Behavioral and perception geography: a critical appraisal,' *Annals of the Association of American Geographers*, 69, 448–62.

16 Y. Tuan (1974) 'A review of P. Gould and R. White, *Mental Maps*, *Annals of the Association of American Geographers*, 64, 591; R. Johnston (1973) 'Mental maps: an assessment,' in J. Rees and P. Newby (eds) *Symposium on Behavioral Geography*, Middlesex Polytechnic.

17 E. Graham (1976) 'What is a mental map?' *Area*, 8, 259–62.

18 I. Cullen (1976) 'Human geography, regional science and the study of individual behavior,' *Environment and Planning*, A, 8, 400.

19 M. Dacey (1975) 'Comments on papers by Isard and Smith,' *Papers of the Regional Science Association*, 35, 56.

20 Cullen, 'Human geography, regional science and the study of individual behavior' (see note 18 above); Bunting and Guelke, 'Behavioral and perception geography' (see note 15 above).

21 J. Louviere (1975) 'The dimensions of alternatives in spatial choice processes: a comment,' *Geographical Analysis*, 7, 315–33.

22 W. Michelson (1977) 'Review of Golledge and Rushton, *Spatial Choice and Spatial Behavior*,' *Annals of the Association of American Geographers*, 67, 603–4.

23 J. Pipkin (forthcoming 1981) 'The concept of choice and cognitive explanations of spatial behavior,' *Economic Geography*.

24 G. Rushton (1969) 'Analysis of spatial behavior by revealed space preference,' *Annals of the Association of American Geographers*, 59, 391–400.

25 R. G. Golledge and G. Zannaras (1973) 'Cognitive approaches to the analysis of human spatial behaviors,' in W. H. Ittelson (ed.) *Environment and Cognition*, New York, Seminar Press, 59–94.

26 For example, Downs and Stea, *Maps in Minds* (see note 12 above).

27 K. O'Connor (1980) 'The analysis of journey-to-work patterns in human geography,' *Progress in Human Geography*, 4, 475–99.

28 See, for example, S. Lieber (1977) 'Attitudes and revealed behavior: a case study,' *Professional Geographer*, 29, 53–8; J. Louviere, 'Information-processing theory and functional form in spatial behavior,' in Golledge and Rushton (eds) *Spatial Choice and Spatial Behavior* (see note 11 above).

29 Pipkin, 'The concept of choice' (see note 23 above).

30 D. Huff (1963) 'A probability analysis of shopping center trade areas,' *Land Economics*, 39, 81–90; W. Baumol and E. Ide (1956) 'Variety in retailing,' *Management Science*, 3, 93–101.

31 R. Downs (1970) 'The cognitive structure of an urban shopping centre,' *Environment and Behavior*, 2, 13–39.

32 P. Burnett (1973) 'The dimensions of alternatives in spatial-choice processes,' *Geographical Analysis*, 5, 181–204.

33 M. Cadwallader (1975) 'A behavioral model of consumer spatial decision making,' *Economic Geography*, 51, 339–49.

34 H. Schuler (1979) 'A disaggregate store-choice model of spatial decision making,' *Professional Geographer*, 31, 146–56.

35 Lieber, 'Attitudes and revealed behavior' (see note 28 above).

36 N. Patricios (1978) 'Consumer images of spatial choice and the planning of shopping centers,' *South African Geographic Journal*, 60, 103–20.

37 See, for example, R. Hudson (1974) 'Images of the retail environment: an example of the use of the repertory grid methodology,' *Environment and Behavior*, 6, 470–90.

38 A. Spencer (1980) 'Cognition and shopping choice: a multidimensional scaling approach,' *Environment and Planning*, A, 12, 1235–51.

39 F. Koppelman and J. Hauser (1978) 'Destination choice behavior for nongrocery shopping trips,' *Transportation Research Record*, 673, 157–65.

40 P. McCarthy (1980) 'A study of the importance of generalized attributes in shopping choice behavior,' *Environment and Planning*, A, 12, 1269–86.

41 See, for example, Downs, 'The cognitive structure' (see note 31 above); and Patricios, 'Consumer images of spatial choice' (see note 36 above).

42 For example, J. Harrison and P. Sarre (1975) 'Personal construct theory in the measurement of environmental images,' *Environment and Behavior*, 7, 3–58; R. Downs, 'Personal constructions of personal construct theory,' in Moore and Golledge, *Environmental Knowing* (see note 13 above).

43 See, for example, Spencer, 'Cognition and shopping choice' (see note 38 above).

44 Wohlwill, J. F. (1976) 'Searching for the environment in environmental cognition research: a commentary on a research strategy,' in G. T. Moore and R. G. Golledge, *Environmental Knowing*, 385–92 (see note 13 above).

45 See, for example, R. Briggs, 'Urban cognitive distance,' in Downs and Stea *Image and Environment* (see note 12 above); R. Briggs, 'Methodologies for the measurement of cognitive distance,' in Moore and Golledge, *Environmental Knowing* (see note 13 above); and R. Lowrey (1970) 'Distance concepts of urban residents,' *Environment and Behavior*, 2, 57–73.

46 M. Cadwallader, 'Cognitive distance in intraurban space,' in Moore and Golledge (eds) *Environmental Knowing* (see note 13 above).

47 Schuler, 'A disaggregative store-choice model' (see note 34 above).

48 M. Pacione (1976) 'Measures of the attraction factor: a possible alter-

native,' *Area*, 6, 279–82.

49 Patricios, 'Consumer images of spatial choice' (see note 36 above).

50 Spencer, 'Cognition and shopping choice' (see note 38 above).

51 P. Stopher (1977) 'On the application of psychological measurement techniques to travel demand estimation,' *Environment and Behavior*, 9, 67–80; G. Nicolaidis (1977) 'Psychometric techniques in transportation planning: two examples,' *Environment and Behavior*, 9, 459–86.

52 Burnett, 'The dimensions of alternatives in spatial choice processes,' 187 (see note 32 above).

53 T. Domenich and D. McFadden (1975) *Urban Travel Demand: A Behavioral Analysis*, Amsterdam, North Holland; E. Sheppard (1979) 'Notes on spatial interaction,' *Professional Geographer*, 31, 8–15.

54 J. Pipkin (1979) 'Problems in the psychological modeling of revealed destination choice,' in S. Gale and G. Olsson (eds) *Philosophy in Geography*, Dordrecht and Boston, Reidel, 309–28.

55 See, for example, Davies, 'Effects of consumer income differences on shopping movement behavior' (see note 4 above); P. Ambrose (1968) 'An analysis of intraurban shopping patterns,' *Town Planning Review*, 39, 327–34; R. Schiller (1977) 'The measurement of attractiveness of shopping centers to middle-class luxury consumers,' *Regional Studies*, 6, 291–7; and C. Thomas (1974) 'The effects of social class and car ownership on intraurban shopping behavior,' *Cambria*, 1, 98–126.

56 See, for example, J. Adams (1969) 'Directional bias in intraurban migration,' *Economic Geography*, 45, 302–23; and R. Johnston (1972) 'Activity spaces and residential preferences: some tests of the hypothesis of sectoral mental maps,' *Economic Geography*, 48, 199–211.

57 R. Potter (1976) 'Directional bias within the usage and perceptual fields of urban consumers,' *Psychological Reports*, 38, 988–90; Potter, 'Spatial patterns of consumer behavior and perception in relation to the social-class variable', (see note 4 above); Potter, 'The nature of consumer usage fields in an urban environment' (see note 4 above).

58 R. Lloyd and D. Jennings (1978) 'Shopping behavior and income: comparisons in an urban environment,' *Economic Geography*, 54, 157–67.

59 S. Taylor (1979) 'Personal dispositions and human spatial behavior,' *Economic Geography*, 55, 184–95.

60 See, for example, K. Craik (1975) 'The personality research paradigm in environmental psychology,' in S. Wapner, S. Cohen and B. Kaplan (eds) *Experiencing the Environment*, New York, Plenum.

61 J. Louviere, L. Ostresh, D. Hensley and R. Meyer (1976) 'Travel demand segmentation: some theoretical considerations related to behavioral modeling,' in P. Stopher and A. Meyburg (eds) *Behavioral Travel-Demand Models*, Lexington, Mass., Lexington Books, 259–70.

62 R. Dobson and M. Tischer (1978) 'Perceptual market segmentation technique for transportation analysis,' *Transportation Research Record*, 673, 145–52.

63 P. Stopher and G. Ergun (1979) 'Population segmentation in urban recreation choices,' *Transportation Research Record*, 728, 59–65.

64 R. Luce and P. Suppes (1965) 'Preference, utility and subjective probability,' in R. Luce, R. Bush and E. Galanter (eds) *Handbook of Mathematical Psychology, III*, New York, Wiley.

65 Pipkin, 'Problems in the psychological modeling of revealed destination choice' (see note 54 above).

66 See, for example, Rushton, 'Analysis of spatial behavior by revealed space preference' (see note 24 above); Cadwallader, 'A behavioral model of consumer spatial decision making' (see note 33 above); Lieber, 'Attitudes and revealed behavior' (see note 28 above); and Schuler, 'A disaggregate store-choice model' (see note 34 above).

67 See, for example, Domenich and McFadden, *Urban Travel Demand: A Behavioral Analysis* (see note 53 above); Charles River Associates (1976) *Disaggregate Travel Demand Models, Phase I Report*, Boston, Charles River Associates.

68 See, for example, Burnett, 'The dimensions of alternatives in spatial-choice processes (see note 32 above); Louviere, 'The dimensions of alternatives in spatial-choice processes: a comment' (see note 21 above); and Sheppard, 'Notes on spatial interaction (see note 53 above).

69 Sheppard, ibid.; Pipkin, 'Problems in the psychological modeling of revealed destination choice' (see note 54 above).

70 R. Luce (1959) *Individual Choice Behavior*, New York, Wiley.

71 A. Tversky (1972) 'Choice by elimination,' *Journal of Mathematical Psychology*, 9, 341–67; S. Thorson and J. Stever (1974) 'Classes of models for selected axiomatic theories of choice,' *Journal of Mathematical Psychology*, 11, 15–32.

72 B. Morgan (1974) 'On Luce's choice axiom,' *Journal of Mathematical Psychology*, 11, 107–23.

73 J. Steele (1974) 'Limit properties of Luce's choice theory,' *Journal of Mathematical Psychology*, 11, 124–31.

74 Z. Shapira and I. Venezia (1977) 'On aggregation across subjects in analyzing individual choice behavior,' *Journal of Mathematical Psychology*, 16, 60–7.

75 Wrigley, N. (1980) 'Paired-comparison experiments and logit models: a review and illustration of recent developments,' *Environment and Planning*, A, 12, 21–40.

76 P. Stopher and A. Meyburg (eds) (1976) *Behavioral Travel Demand Models*, Lexington, Mass., Lexington Books.

77 See, for example, Stopher 'On the application of psychological measurement techniques' (see note 51 above).

78 For example, Charles River Associates (1978) *New Approaches to Understanding Travel Behavior, Phase II* (two volumes), Boston, Charles River Associates.

79 For example, Burnett, 'The dimensions of alternatives in spatial choice processes' (see note 32 above); and M. Cadwallader (1975) 'A behavioral model of consumer spatial decision making,' *Economic Geography*, 51, 339–49.

80 For example, Lieber, 'Attitudes and revealed behavior (see note 28 above); and Schuler, 'A disaggregate store choice' (see note 34 above).

81 For example, Rushton, 'Analysis of spatial behavior by revealed space preference' (see note 24 above).

82 Pipkin, 'Problems in the psychological modeling of revealed destination choice' (see note 54 above).

83 Bunting and Guelke, 'Behavioral and perception geography,' 455 (see note 15 above).

84 For example, Cadwallader, 'A behavioral model of consumer spatial decision making' (see note 33 above).

85 For example, Patricios, 'Consumer images of spatial choice,' 113 (see note 36 above).

87 Louviere, 'Information-processing theory and functional form in spatial behavior' (see note 28 above); S. Lieber, 'A comparison of metric and nonmetric scaling models in preference research,' in Golledge and Rushton, *Spatial Choice and Spatial Behavior* (see note 11 above).

86 Charles River Associates, *Disaggregate Travel Demand Models* (see note 67 above).

88 Spencer, 'Cognition and shopping choice' (see note 38 above).

89 E. Sheppard (1980) 'The ideology of spatial choice,' *Regional Science Association*, 45, 197–213.

90 For example, Stopher and Meyburg, *Behavioral Travel Demand Models* (see note 61 above); M. Johnson (1978) 'Attribute importance in multiattribute transportation decisions,' *Transportation Research Record*, 673, 15–21; R. Meyer, I. Levin and J. Louviere (1978) 'Functional analysis of mode choice,' *Transportation Research Record*, 673, 603–4; S. Lerman and J. Louviere 91978) 'Using functional measurement to identify the form of utility functions in travel-demand models,' *Transportation Research Record*, 673, 78–86; R. Dobson, F. Dunbar, C. Smith, D. Reibstein and C. Lovelock (1978) 'Structural models for the analysis of traveller attitude behavior relationships,' *Transportation*, 7, 351–63; and D. Hensher and J. Louviere (1979) 'Behavioral intentions as predictors of very specific behavior,' *Transportation*, 8, 167–82.

91 ibid.

92 P. Burnett and D. Ellerman (1981 forthcoming) 'New methodologies for modeling the travel behavior of diverse human groups in American cities,' *Transportation Research Record*; P. Burnett and S. Hanson (1979) 'Rationale for an alternative mathematical approach to movement as complex human behavior,' *Transportation Research Record*, 723, 11–24; S. Hanson (1976) *Urban Travel Linkages: A Review*, Adelaide, Australia, Third International Conference on

Behavioral Modeling, Resource Paper; S. Hanson (1980) 'Spatial diversification and multipurpose travel: implications for choice theory,' *Geographical Analysis*, 12, 245–57; S. Hanson (1980) 'The importance of the multipurpose journey to work in urban travel behavior,' *Transportation*, 9, 229–48; I. Heggie (1978) 'Putting behavior into behavioral models of travel choice,' *Journal of the Operations Research Society*, 29, 541–50; I. Heggie and P. Jones (1978) 'Defining domains for models of travel demand,' *Transportation*, 7, 119–25; P. Jones (1979) ' "HATS": a technique for investigating household decisions,' *Environment and Planning*, A, 11, 59–70; D. Hensher (1979) 'Market segmentation as a basis for allowing for variability in traveller behavior,' *Transportation*, 8, 167–82; T. Van der Hoorn (1979) 'Travel behavior and the total activity pattern,' *Transportation*, 8, 309–28.

93 J. Wheeler (1972) 'Trip purposes and urban activity linkages,' *Annals of the Association of American Geographers*, 62, 641–54.

94 Hanson, *Urban Travel Linkages* (see note 92 above); Hanson, 'The importance of the multipurpose journey to work in urban travel behavior' (see note 92 above).

95 Schuler, 'A disaggregate store-choice model of spatial decision making' (see note 34 above).

96 For example, T. Adler and M. Ben-Akiva (1979) 'A theoretical and empirical model of trip-chaining behavior,' *Transportation Research*, 13b, 243–57.

97 L. Curry (1962) 'The geography of service centers within towns,' *Lund Studies in Geography*, C, 24, 31–53.

98 P. Tranter and D. Parkes (1979) 'Time and images in urban space,' *Area*, 11, 115–20.

99 D. Parkes and N. Thrift (1975) 'Timing space and spacing time,' *Environment and Planning*, A, 7, 651–70; D. Parkes and N. Thrift (1978) 'Putting time in its place,' in T. Carlstein, D. Parkes and N. Thrift (eds) *Timing Space and Spacing Time, I*, London, Arnold.

100 T. Hägerstrand (1974) 'The impact of transport on the quality of life,' in *Transport in the 1980–1990 Decade*, Fifth International Symposium Theory and Practice in Transport Economics, OECD, Paris; B. Lenntorp (1976) *Paths in Time–Space Environments*, Lund, Gleerup.

101 D. Hensher (1976) 'The structure of journeys and the nature of travel patterns,' *Environment and Planning*, 8, 655–72.

102 Heggie and Jones, 'Defining domains for models of travel choice' (see note 92 above).

103 Burnett and Hanson, 'Rationale for an alternative mathematical approach' (see note 92 above).

104 Jones, ' "HATS": a technique for investigating household decisions' (see note 92 above).

105 Burnett and Ellerman, 'New methodologies for modeling the travel behavior of diverse human groups in American cities' (see note 92 above).

106 Burnett and Hanson, 'Rationale for an alternative mathematical

approach' (see note 92 above).

107 Hensher, 'The structure of journeys and the nature of travel patterns' (see note 101 above).

108 Heggie, 'Putting behavior into behavioral models of travel choice' (see note 92 above).

109 I. Cullen (1976) 'Human geography, regional science and the study of individual behavior,' *Environment and Planning*, A, 8, 397–409.

110 Heggie, 'Putting behavior into behavioral models of travel choice,' 541 (see note 92 above).

111 Cox and Golledge, *Behavioral Problems in Geography*, introduction (see note 10 above); D. Harvey (1967) *Behavioral Postulates and the Construction of Theory in Human Geography*, University of Bristol, Department of Geography, Seminar Paper Series A(6).

112 G. H. Bower (1975) 'Cognitive psychology,' in W. Estes (ed.) *Handbook of Learning and Cognitive Processes, I*, Hillsdale, N.J., Erlbaum, 25–80; U. Neisser (1967) *Cognitive Psychology*, New York, Appleton-Century-Crofts.

113 A. Paivio (1971) *Imagery and Verbal Processes*, New York, Holt, Rinehart & Winston.

114 Pipkin, 'The concept of choice and cognitive explanations of spatial behavior' (see note 23 above).

115 For example, L. Perlmuter and R. Monty (1979) *Choice and Perceived Control*, Hillsdale, N.J., Erlbaum.

116 Rushton, 'Analysis of spatial behavior by revealed space preference' (see note 24 above).

117 Golledge and Rushton, *Spatial Choice and Spatial Behavior* (see note 11 above).

118 I. Steiner, 'Three kinds of reported choice,' in Perlmuter and Monty, *Choice and Perceived Control* (see note 115 above).

119 For example, Burnett, 'The dimensions of alternatives in spatial-choice processes' (see note 32 above); Cadwallader, 'A behavioral model of consumer-spatial decision making' (see note 79 above); Schuler, 'A disaggregate store-choice model of spatial decision making' (see note 34 above); and Patricios, 'Consumer images of spatial choice' (see note 36 above).

120 R. Golledge *et al.*, 'Learning about a city: analysis by multi-dimensional scaling,' in Golledge and Rushton, *Spatial Choice and Spatial Behavior* (see note 11 above).

121 For example, T. Herzog, S. Kaplan and R. Kaplan (1976) 'The prediction of preference for familiar urban places,' *Environment and Behavior*, 8, 627–55; Tranter and Parkes, 'Time and images in urban space' (see note 98 above).

122 Paivio, *Imagery and Verbal Processes* (see note 113 above); A. Paivio (1976) 'Imagery in recall and recognition,' in J. Brown (ed.) *Recall and Recognition*, New York, Wiley, 103–30.

123 Paivio, *Imagery and Verbal Processes*, 9 (see note 113 above).

124 S. Kaplan, 'Cognitive maps in perception and thought,' in Downs and Stea, *Image and Environment* (see note 12 above).

125 Moore and Golledge, 'Environmental knowing: concepts and theories,' 6 (see note 13 above).

126 R. Beck and D. Wood (1976) 'Cognitive transformation of information from urban geographic fields to mental maps,' *Environment and Behavior*, 8, 199–238.

127 Brown, *Recall and Recognition* (see note 122 above).

128 Harvey, *Behavioral Postulates and the Construction of Theory* (see note 111 above).

129 Cox and Golledge, *Behavioral Problems in Geography* (see note 10 above).

130 Harvey, *Behavioral Postulates and the Construction of Theory*, 24 (see note 111 above).

131 Rushton, 'Analysis of spatial behavior,' 49 (see note 24 above).

132 Golledge, R. G. (1979) 'Reality, process and the dialectical relation between man and environment,' in S. Gale and G. Olsson (eds) *Philosophy in Geography*, 109–20 (see note 54 above).

133 For example, H. Simon (1979) *Models of Thought*, New Haven, Yale University Press.

134 B. Skinner (1953) *Science and Human Behavior*, New York, Free Press.

135 Burnett and Hanson, 'Rationale for an alternative mathematical approach' (see note 92 above).

136 G. Olsson, 'Inference problems in locational analysis,' in Cox and Golledge, *Behavioral Problems in Geography* (see note 10 above); L. Svart (1974) 'On the priority of behavior in behavioral research: a dissenting view,' *Area*, 6, 301–4.

137 ibid., 303.

138 Cadwallader, 'A behavioral model of consumer spatial decision making' (see note 79 above).

139 Burnett and Hanson, 'Rationale for an alternative mathematical approach' (see note 92 above); Sheppard, 'The ideology of spatial choice' (see note 89 above); Pipkin, 'The concept of choice' (see note 23 above).

140 Smith, 'An axiomatic theory of spatial discounting behavior' (see note 8 above); Smith, 'A choice theory of spatial interaction' (see note 8 above).

141 Luce, *Individual Choice Behavior* (see note 70 above).

142 Hanson, 'The importance of the multipurpose journey to work' (see note 92 above).

143 For example, J. Harrison and W. Howard (1972) 'The role of meaning in the urban image,' *Environment and Behavior*, 4, 389–411.

144 S. Rosenberg and H. Simon (1977) 'Modeling semantic memory: effects of presenting semantic information in different modalities,' *Cognitive Psychology*, 9, 293–325.

145 Y. Tuan (1975) 'Images and mental maps,' *Annals of the Association of American Geographers*, 65, 205–13.

146 Cullen, 'Human geography, regional science and the study of individual behavior' (see note 18 above).

147 F. Cesario and T. Smith (1975) 'Directions for future research in spatial interaction modeling,' *Papers of the Regional Science Association*, 35, 57–72.

148 But see H. Blommestein, P. Nijkamp and W. Veenendaal (1980) 'Shopping perceptions and preferences: a multidimensional attractiveness analysis of consumer and entrepreneurial attitudes,' *Economic Geography*, 56, 167–74.

149 For example, B. Lentnek, S. Lieber and I. Sheskin (1976) 'Consumer behavior in different areas,' *Annals of the Association of American Geographers*, 66, 538–45.

8 Residential mobility and behavioral geography: parallelism or interdependence?

W. A. V. Clark

In any review and discussion of the role and influence of the behavioral methodology on geography in general and on specific substantive fields within geography, we are necessarily faced with the problem of trying to assess the extent to which there is a discernible impact of one on the other, and whether there is an accidental connection or simply a parallelism between trends in geography and in other social sciences. Specifically, this chapter is designed to examine the degree of parallelism and/or interdependence between behavioral geography and residential mobility. Have they influenced one another? What are the common threads which link these two diverse fields?

There is no doubt that there has been an explosion in the literature on residential mobility in the last dozen years nor is there any doubt that this parallels the development of behavioral geography. The increasing tempo of research in behavioral methodology and residential mobility dates from the last few years of the 1960s, the same time as initial work on behavioral geography was published.[1] But to examine the question of parallelism versus interdependence, we need a framework within which to organize the review and discussion.

Residential mobility can be described as a truly interdisciplinary field that is focused on population relocation within the city, and the

I would like to thank my colleague, Nick Entrikin, and the editors for comments on earlier drafts of this paper.

causes, implications and related policy issues of those relocations. The issues raised by population relocation within the city have been examined from almost all disciplinary perspectives, but particularly by sociologists, who have been interested in the demographic influences on mobility, by economists, who have focused on the links between mobility and housing markets, by geographers, who have been concerned with an overall analysis of the changing spatial structure of the city in relationship to residential mobility, and by historians, who have considered the implications of residential mobility on the formation of ethnic areas within the city.

Within the substantive field of residential mobility we can identify a major general trend that has parallels in the development of the behavioral methodology. There has been an increasing concern with an individual approach to residential mobility, which developed in response to certain perceived inadequacies in the more macro-analytic or aggregate studies of mobility. There has been a continuing trend towards the analysis of the individual migrant and his decision-making behavior. Such studies did not develop suddenly; in fact, the first and seminal work to incorporate a behavioral approach to residential mobility is Rossi's study, *Why Families Move.*[2] As a sociologist, he was concerned with a variety of broad issues relating to individuals, neighborhoods and the adjustments that households make as they change their social and physical space. Similarly, studies by Goldstein,[3] Myers, McGinnis and Pilger[4] and Wolpert[5] also emphasized an individual and behavioral focus. These studies developed in response to the lack of detail that could be derived from aggregate-flow studies between origins and destinations. For some social scientists a simple focus on individual rather than aggregate patterns was sufficient motivation. For others, the study of the individual migrant as distinct from aggregate flows reflected a concern with the behavior of the individual migrant. To these social scientists the emphasis was not simply to understand individual flows and patterns, but rather to understand individual decision making in residential mobility within a social psychological context, and thus better understand the way in which individuals make choices within the city. It is important to differentiate the simple focus on individual migrants from the emphasis on the social psychological themes that might be used as 'true explanations' of mobility behavior.

Similar trends and concerns within geography can be identified as crucial in the development of behavioral methodology and behavioral geography. There had always been an interest in spatial form or spatial structure, but a process or dynamic approach to the formulation of spatial structure was lacking. The behavioral methodology has in it the same trends that we have observed in the substantive field of residential mobility. There was a concern to focus on the individual, be it the shopper or the migrant, and to observe his behavior in space, rather than to examine the patterns of shopping centers, of land uses and of the structural elements of the city. Secondly, there was a concern to develop explanations of these spatial behaviors, which led to social psychological concerns and attempts to use mental maps and cognitive processes to bring understanding to spatial processes.

From the beginning, individuals writing on, and having a concern for, behavioral geography distinguished between the study of spatial structure, the behavior of those structures and behavior within those structures.[6] While there was a substantial array of descriptive and analytic tools that were concerned with the study of spatial form or spatial structure, methods for the study of behavior within structures and of the processes of structural change were much more limited. Certainly, the concern with a more process-oriented geography[7] was a driving force in the development of behavioral geography. It is the focus on process that is the most general and most basic element of behavioral geography.

The process orientation suggested a more detailed analysis of individual behavior, which in turn was partitioned into concerns with behavior in space (the dynamic process of people functioning within a structure) and spatial behavior (the rules (yet to be discovered) by which people made decisions and carried through their behaviors in space).[8] This division set up a context for the search for a set of structured relationships, theories and laws that were intended to organize the way in which individuals conceptualized and responded to the spatial structure.

A concern with understanding an individual's behavior (the choices, decision making and spatial outcomes) was further stimulated by the view that there had been an overly long focus on classical economic man. A redirected focus on non-optimal human behavior, initially from social psychology,[9] emphasized the effects of habit (or role playing), of social pressures and of cultural institutions and the influence these have on preferences and choice, and hence the outcomes of individual behavior in a spatial setting. As Burnett notes, the emphasis is not just on understanding the more complex set of variables, including economic measures, but also on recognizing the importance of a longitudinal or truly process approach to understanding behavior within the city.[10]

The initial simple concern with process and observable human behavior was followed by an increased concern to understand the way in which human behavior was influenced by people's perception of the environment and by a concern for decision making within the spatial context. The focus on environmental cognition or spatial cognition has dominated recent research in behavioral geography and has emphasized the way in which the mind forms images and utilizes these images in actual behavior. A variety of studies have attempted to develop procedures 'for recovering and measuring perceptions and preferences and relating them to overt spatial choice behavior.'[11]

It is clear that behavioral geography and a specific concern with human behavior in a spatial structure is now of some significance in geographical research and teaching. In a recent text book King and Golledge summarize the concerns of behavioral investigations. They note that:

> The relationships between . . . people and the environment is influenced by the spatial form of the city . . . and in turn the spatial form of the city is formed by human behavior . . . . By studying patterns of human behavior in the city, by seeking to comprehend the different mental pictures that people have of the city and its parts and by probing into how these pictures develop and change we can begin to appreciate the complex interaction between the spatial form of the city and the patterns of human behavior.[12]

Following these broad introductory remarks on the substantive area of residential mobility and the behavioral methodology, we can set about an examination of the extent to which they have paralleled or influenced one another.

Review and interpretation

Given the large literature that focuses on residential mobility, this review selects those materials that seem to have had the most influence on the changing trends in research on mobility.[13]

Micro–macro framework of analysis

Whereas, once, residential mobility was a poor relation of studies of migration, it now stands as an equal area of investigation within the general field of population redistribution and change. To some extent the differences between the explanations that have been used for state-to-state or region-to-region moves and population change within cities have blurred.

The concern now is to develop an integrated approach to migration and mobility in which theory will link these two distinct levels of analysis. This merging of the approaches to understanding residential relocation has emphasized the significance of what have been called the micro-analytic or behavioral approaches to residential mobility. Within both aggregate analyses of migration flows between origins and destinations and micro-analyses of relocation decision making within the city there is increasing concern with 'behavioral' explanations. There is still a distinction between micro and macro approaches. Macro approaches continue to deal with flow matrices as well as origin and destination sets, while micro approaches attempt to elucidate decision-making behavior. However, the macro-analytic and behavioral studies are properly viewed as complementary, and yield insights into each other as well as into residential mobility as a whole. For example, knowing why an individual decides to move, and the impact on the system of his decision to change residences, can lead us to a better understanding of why some districts are areas of high in- and out-migration and why some areas are more stable. While the micro-analytic or behavioral approach focuses on the decision to move, on the role of differential access to sources of information in shaping that decision and on the spatial patterns of search, the macro-analytic is still largely concerned with the spatial regularities in migration streams and the interrelationship between areas of in- and out-migration.[14]

At the pragmatic level, use of aggregate data and a macro-analytic approach reduces the complexity of analysis and obviates the need to obtain the large number of detailed observations required for parameter estimation in models of individual decision making. The best summary statement is to say that each mode of analysis can provide information which is relevant to the other and both can yield insights into residential mobility.

An evaluation of the mobility literature

We have identified four studies that were significant in impacting the direction of research in residential mobility.[15] The thrust of these studies was to shift the emphasis from aggregate-pattern analysis, which was largely descriptive and interpretative, to studies that focused on the individual household, emphasized the modeling of the interrelationships that make up residential mobility patterns, and emphasized concepts drawn from psychology and social psychology. It is not necessary to review these well-known studies in detail, but it is important to note that Rossi's investigation was concerned with both individual households and areas. It was truly a transition study from the earlier focus on origins and destinations and the formation of neighborhoods, to an emphasis on the characteristics that distinguish mobile and stable households, and to the role of the family life cycle in creating mobility.

Two studies in the same decade as the Rossi study were also important in the development of behavioral thinking in mobility research. The work initiated by Goldstein in the Norristown Study emphasized the use of individual longitudinal records.[16] While the central questions at issue in the study were analyses of the changing composition and demographic behavior of the population specifically with respect to the affect of occupational mobility on labor force composition, a great deal of new conceptual thinking was developed from the research. Goldstein was able to investigate the nature of repeated migration and the distinction between 'movers' and 'stayers.' The analysis of longitudinal data files enabled the first detailed discussion of migration histories and the behavior of specific groups of migrants as they adjusted to specific locations in the urban environment. The implications of these studies spilled over into questions of stability, social organization and neighborhood structure. The focus on and concern with individual data files had two specific results for studies of mobility and in turn provide a connection to behavioral processes. The acceptance of individual data raised the question of the level of understanding being provided by these studies. The individual records raised the possibility of more detailed explanations, of explanations that involve beliefs and aspirations of the actual processes of decision making in residential mobility. Of these studies it was Rossi's emphasis on the family life cycle that moved analyses of residential mobility away from general-pattern analyses towards an understanding of the mobility process itself. It remained for Myers, McGinnis and Pilger, and Wolpert to shift the emphasis from description to conceptual formulation.[17]

Even though, as Wolpert himself notes, 'The model . . . is of doubtful usefulness as an exact predictive tool . . ., it borrows much of its concepts and terminology from the *behavioral theorists* because of the intuitive relevance of their findings to the analysis of mobility.'[18] Clearly Wolpert moved the analysis from the descriptive and from spatial interaction to the behavioral and the dynamic, and was one of the first in geography (or the other social sciences) to turn to the social psychological literature for conceptual structures for spatial analysis. The fourth group of articles that can be identified as having an important impact on the development of thinking in residential mobility research, is the specific attempt to *model* social mobility and, by implication, residential mobility.[19] The 'Cornell Mobility Model' is organized around the axiom of cumulative inertia – which states that the probability of remaining in any state of nature (or in any physical location) increases as a strict monotonic function of duration of prior residence in that state. While the actual modeling procedure utilized a Markov approach, and, in fact, used aggregate data, the importance of the conceptual thinking qualifies this research as an important element of work that is increasingly concerned with understanding residential mobility. It is also significant in that it is one of the first attempts to lay out a formal structure for social and residential mobility.

To a large extent, these books and articles defined the research themes in demography, sociology, economics and geography. From Rossi the investigations of mobility attempted to link the probability of moving to specific events in the life cycle and gathered national surveys of moving behavior.[20] From Goldstein and McGinnis a variety of studies have attempted formal modeling of the time-dependent nature of mobility, as emphasized in the work of Ginsberg and summarized in Boudon.[21] From Wolpert, the direction has been fragmented into investigations of stress, studies of mental maps and perception, and concerns with information processing.[22] As yet there is no adequate summary of this latter material, although Speare, Goldstein and Frey utilize insights from Wolpert, Adams, and Brown and Moore.[23]

The themes that have developed in response to the identification of life-cycle concepts and the time dependency of residential mobility have either considered specific links between life events and housing or the statistical properties of the mobility process.

Many of the studies that analyzed the life-cycle approach to understanding mobility did so within a survey framework in which the investigations considered the relative weights of life cycle versus other events that might explain mobility. Certainly the role of the life cycle was an integral element of Wolpert's attempt to formalize mobility in terms of stress and of Butler, Sabagh and Van Arsdol's concern with the role of dissatisfaction in residential mobility.[24] More recently Speare, Goldstein and Frey considered specific life events in their attempt to model the role of stress in mobility. This latter work, however, explicitly considered the role of housing and neighborhood variables and attempted in that way to set the behavior within a broader urban context. It is the role of housing in mobility that has stimulated the investigations by urban economists who have suggested that up to the present 'the theories (of residential mobility) provide little in the way of specific hypotheses or verifiable propositions.'[25]

Although a series of explicit models of intraurban mobility have been developed, the extent to which the models are successful still needs to be evaluated.[26] The models seem to grow out of a specific concern with changes in the housing market rather than an interest in residential mobility *per se*. In particular, mobility is seen as a response to changes in the demand for housing services. In any event, the Quigley and Hanushek view of residential mobility is closely related to that initially expressed by Rossi, even though they have framed it in terms of utility maximization. Quigley and Weinberg note that mobility will be more likely to occur if 'the dollar value of the benefits derived by moving to a new dwelling unit exceed the cost associated with that move.'[27]

Whether or not these models will survive vigorous empirical testing is unimportant to this discussion. The important issue is that the models have been formulated in response to geographic models emphasizing attitude formation and attitude response – models that have neglected the economic influences on household behavior. In most of the sociological and demographic models, economic variables have been subsumed within the more general concerns for stress and satisfaction and there has been no attempt at cost–benefit analysis. However, the solidly behavioral focus of these economic models is emphasized in the comment that traditional economic analysis of urban housing markets[28] masks much of the disequilibrium and the incentives for adjustment by averaging across those individuals who are underconsuming and those who are over-consuming housing.[29]

Studies of the time-dependent nature of residential moves have looked at population changes in which duration of residence or the decision *not* to move is the focus. While the axiom of cumulative inertia is widely cited as an explanation of residential mobility, in fact it is an explanation of why people do not move. The explanation embodied in the axiom of cumulative inertia embodies an hypothesis of the way in which people become attached to a neighborhood and to a particular house or apartment through acquaintances, acquisitions and memories. These ties to the local area become stronger over time and more difficult to break. As yet there is no full explanation of the role of inertia in residential decision making. In the decade since the introduction of the axiom of cumulative inertia most of the research has focused not on the behavioral aspects of the axiom, but rather on the statistical properties of time-dependent mobility. Research on the independent-trials process has been designed to argue that the observed decline in mobility rates can be accounted for by a population in which there is heterogeneity.[30] Other approaches, particularly those by Ginsberg put forward in a series of articles,[31] suggests that the McGinnis model can be expressed as a 'semi-Markov' or 'Markov-renewal' process. In particular the work by Spilerman and McGinnis has expanded the methodologies for the analysis of residential mobility, but it has emphasized the statistical analysis of the probabilities of moving at the expense of both explanatory modeling of the processes of mobility and the role of conceptual variables in explaining the decision-making and search processes involved in residential relocation.

The geographic work is more fragmented than that by demo-graphers and economists.[32] Wolpert's work led to attempts to specify the concepts of place utility, search space and action space. The thrust of Brown and Moore's article[33] was to provide a more elaborate structure for the concept of stress and stress response. Their particular contribution was to emphasize the notion of a dissatisfaction or stress level that initiates the decision to seek a new residence. Additionally the formulation emphasized the distinction between the decision to move and the actual relocation. Several studies have focused on place utility[34] and on the way in which stress is generated.[35]

As we have already noted, the focus on stress (or dissatisfaction) has also been incorporated into work by demographers and economists. But while it implicitly underlies the investigations that are now ongoing in urban economics there is no direct parallelism between the work by demographers, sociologists, geographers and economists. The common thread amongst geographers and demographers involves the households' response to the degree of dissatisfaction (or stress) that is impinging on the household. Speare, Goldstein and Frey[36] and Huff and Clark[37] develop theoretical descriptions in which they view mobility as the results of a constant decision-making process in which the basic decision to move is related to an increase in dissatisfaction beyond some threshold or tolerance level. Economists relate stress or dissatisfaction (and mobility) to disequilibrium in housing consumption.

The other approaches to residential mobility stimulated by Wolpert's work – which are often considered to be the most 'behavioral' – while promising a great deal, have in fact made only minor contributions. Although discussions of mental maps have been related to preferences, research on mental maps has, to a large extent, been 'mental mapping' for its own sake.[38] The few attempts to integrate and use mental maps as a procedure for understanding residential preferences, and thereby for understanding people's residential behavior, have failed to specify an adequate link between preference and behavior. Some investigations have examined bias while others have focused on the information-processing aspects of the mental map.

Recent thinking on the role of mental maps in understanding spatial behavior has suggested that 'cognitive maps play only a minor and intermittent role in effective thinking, and that it is misleading to impute them any great significance in the coordination of our spatial activities.'[39] While behavior may be governed by 'some kind of cognitive structure' we are a considerable distance from understanding that structure. The alternative to simple 'mental mapping' and the investigation of cognitive structures may lie in the investigation of risk aversion, belief systems and information updating as a means of uncovering what underlies spatial behavior.[40]

Almost all the attempts to integrate the mental-map approach with residential mobility have focused on a variety of aspects of directional bias. Adams has suggested that each individual has a mental image of the city that is sectoral in form, and which emphasizes the portion of the city in which s/he lives.[41] Because the journey to work traverses a particular sector of the city, the individual develops a well-structured and more complete knowledge of that wedge-shaped area of the city. There have been a variety of attempts to investigate this mental concept of the city, with varying results. Clark, Donaldson and Johnston, amongst others, have studied the way in which migrants may have biased their search space on the basis of their mental maps.[42] However, almost all of this work has been inferential. There have been very little data gathered that both analyze the movements of individuals and utilize the actual mental maps of those same individuals.

The studies of revealed preference and its extension to information integration has had a more limited impact on studies of residential mobility. The revealed-preference models developed for the examination of consumer behavior,[43] when applied to mobility data,[44] have yielded results that are difficult to interpret. It is not clear whether this is a function of the revealed-preference methodology or of the difficulty of identifying variables that are useful surrogates in trade-off analyses. While distance from center and size of center have been identified as useful trade-off variables in the analysis of consumer behavior, there are no obvious two-dimensional sets that can be used for the analysis of intraurban migration. An additional issue identified by Shepherd may be even more important.[45] Do the revealed preferences of individuals, as they move in the city, in fact tell us very much about the choices and needs that people wish to satisfy? In fact Shepherd's critique serves to emphasize the distinction between the preference work, which is closer to micro-economic analysis, and the behavioral focus on cognitive processes. The revealed-preference analysis cannot by definition tell us how or why the decision was made. Observed behavior may not be a very useful method for finding out about an individual's preferences. While individuals in the city may have a preference for new housing in the suburbs when there is new housing being constructed, it is not clear what happens to those preferences in the absence of opportunities. Still, at the present time, the revealed-preference methodology has not been an important component of behavioral investigations of residential mobility.

Recent reports and analyses of residential mobility, especially of the behavioral approach, have taken a critical perspective. The central theme of analyses by Short, and Bassett and Short, is that the research on residential mobility of the private housing market has emphasized the individual-behavioral aspects of residential mobility at the expense of the wider context within which residential mobility occurs.[46] Some of this critical perspective is drawn from marxist ideology, while other elements seem to be the result of an evaluation of residential mobility in the more constrained markets that exist for many European cities. These elements are not, however, mutually exclusive.

The particular perspective of these critiques is that the behavioral studies that emphasize choice and decision making have exclusively focused on demand factors in the housing system, while the constraints and opportunities within the housing system (the supply side) have not been given equal emphasis. These commentaries serve the important purpose of emphasizing that residential mobility takes place within a broader social and economic system. At the present time, there has been little new research that shows the way in which this constraints process or differential access to housing can be built into analytic studies of residential mobility. To some extent, the studies have directed attention away from the process of mobility and towards the structure of housing markets and their social systems. But, as Maher notes:

> the differing emphasis in the behavioral matrix between choice and constraint is a question of scale. By assuming away the societal and institutional context, most studies have tended to concentrate on the decision framework and the independent variables explaining mobility patterns. At a higher level of abstraction . . . the change is . . . from a variable with independent status to one which is seen to be dependent on a further set of factors.[47]

The question of balance between micro-analytic models and institutional constraints studies is still open.

Review and critique: behavior and the role of information

At this point we can identify a major division in the approaches within 'behavioral geography.' Some behavioral geographers have focused on cognitive processes and problems of spatial choice and the mental processing that underlies individual decision making. Such approaches have been designed in an attempt to understand the processes responsible for behavior acts,[48] even though these approaches have been criticized as being intent on delving 'deeper into the human psyche.'[49] Alternatively other behavioral investigations have been content with studies of perception, information surfaces and revealed preference. Those approaches that have discussed behavioral theory in a more general context have been less successful, and have usually involved the adoption of methods from psychology for the development of a 'general theory of spatial behavior.'[50] This second characterization best describes the behavioral approach to mobility in the past, while the former, directed as it is to substantive problems and couched in terms of decision-making processes, has the greatest unrealized potential for behavioral analyses. The thrust of research in residential mobility is not, however, behavioral in the sense that it will fit within a theory of behavioral geography but is so, rather, in the sense that it is concerned with *individual, longitudinal studies of decision making in a complex environment*. The way in which spatial choices are made is of fundamental concern to anyone investigating residential mobility, but the issues of mental maps, images and preference are not the best means of achieving this objective.

For example, while it is a fairly simple procedure to collect information on the personal attributes of individuals, the composition and distribution of housing within the city, and the preferences of individuals, it is much more difficult to get at the beliefs that individuals have about the housing market. Yet the parameters relating to certainty, reliability and bias are crucial in understanding people's sequential viewing of vacancies within the city. How an individual views a particular house, his own or another, is a complex problem involving a multidimensional data set. At best, in an empirical study, it is only possible to get meaningful information on four or five variables. Thus, for example, in examining the preferences of individuals as to the size and quality of dwelling and the quality of neighborhood they prefer, the actual measurable indices for these complex variables may be as simple as number of bedrooms, type of construction and the amount of green space and trees.[51] Obviously this is a very imperfect subset of the complex of measures that impact on an individual when he views a residence as a potential home. The problems of understanding and explanation are further complicated by the necessity of evaluating the way in which this information is updated (over time) as the individual conducts the search process. Some form of diary or longitudinal follow-up is essential for even the most minimal information on the search procedure.

The concern to model and understand decision making, to understand the behavior that is residential mobility, has increasingly recognized the central role of information in motivating behavior. That information is one of the significant explanatory components of decision making on different scales, and especially in most aspects of consumer choice and behavior, is well recognized in the literature in psychology, economics and marketing, and more recently in geography.[52] There is no disagreement that studies of information are fundamental to a better understanding of decision making and of the role of uncertainty in individual consumer behavior. However, within geography, discussion of information and information use has been more general and descriptive. Information has been an implicit, not explicit, variable in the models. Such studies can be organized into: studies of mental maps and the cognition of the urban environment,[53] attempts to model the diffusion of innovations,[54] studies of learning[55] and analyses of spatial choice.[56]

Only recently have there been specific attempts to evaluate and model information use in residential decision making. Amongst these approaches Louviere has utilized the methodology of human information processing to study spatial behavior in the housing market.[57] Huff and Clark, and Speare, Goldstein and Frey, have attempted to incorporate information and information use, via stress thresholds, into models of residential decision making.[58]

Whether it is the work of sociologists (e.g. Speare, Goldstein and Frey), economists (e.g. Hanushek and Quigley) or geographers (e.g. Huff and Clark),[59] there has been a common concern with the relationship between the decision to move and the motivation of dissatisfaction or 'locational stress.' Information is imbedded at least implicitly in the development of locational stress.

In the Huff and Clark study, a scenario is developed whereby a move will cause a reduction in the stress experienced prior to the move, but, at the same time, the move will dissolve ties with the previous location and so reduce the cumulative inertia. The behavioral component is focused on the tension within each individual, or individual household, regarding the selection of a new residence. On the one hand, there is a resistance to moving, which is engendered by emotional and physical ties to the present residence, coupled with the monetary and psychic costs associated with the actual move. But, on the other hand, the individual may be dissatisfied with attributes of the present dwelling or environment. The model is constructed around the notion that the probability of moving will be a function of the resultant of these conflicting forces. Similar notions are expressed in the Speare, Goldstein and Frey's models, and also by Hanushek and Quigley, who focus on the nature of dissatisfactions with the specific consumption of housing. In each case the approaches are concerned with longitudinal frameworks, with specific modeling of decision making, and they focus on the individual household and its interaction with the urban environment. The studies are behavioral within the pragmatic definition we have already discussed.

The analyses that have focused on the spatial aspects of residential relocation have also had a longitudinal, individual, and even more specific information component. Although there have been a variety of attempts to model residential search (certainly one of the most critical questions in terms of spatial behavior and decision making), there are as yet no pragmatic and acceptable models of search behavior. Geographers, economists and sociologists have drawn from the general search literature that has been developed in economics, which is focused on a variety of aspects of optimal and optional stopping rules. Geographers have taken one direction, economists another, and there are a variety of other studies that are not organized around any specific modeling strategy, although all recognize the significant role of information acquisition and use.

One approach couches the search for a new home in a utility maximization framework: 'The expected utility attached to search for a dwelling in a given neighborhood is based upon the household's prior estimates of the housing characteristics in the area . . . and the expected costs of search in that area.'[60] If the expected utility of search in a given area is greater than the household's present utility then it is assumed that the area is included within the search space of the prospective migrant. A comparison of expected and actual utility leads to the decision to search, and search is initiated in the neighborhood that has the greatest difference between the expected and actual utility.

The other extensive study of search behavior utilizes the questionnaire results of a sample of the population who engaged in the Housing Allowance Demand Experiment.[61] They were able to establish a number of what we can term sociological observations. A household's decision to search for housing is associated with certain household characteristics including the age of head of household, previous mobility, and the degree of satisfaction with the housing unit and the neighborhood. Such results were earlier reported by Speare, Goldstein and Frey.[62] The logit models reported in the study are designed to predict which households are more likely to search.[63]

Investigations of search behavior have demonstrated the interconnection between traditional economic, demographic and geographic approaches and methodologies, and concepts drawn from psychology and other work in behavioral geography. The research is focused on a continually changing process, includes a range of traditional variables, and has added measures of attitudes and beliefs. The motivating force is not the need to develop behavioral theory, but the need to understand how people relocate within the city. While the research by economists and sociologists has emphasized traditional regression models, geographers have attempted to utilize decision theory as a component of their explanations of spatial search, although as yet geographers have not been able to translate their descriptive discussions of information use to formal models. This latter step is crucial if geographers are to influence the general work in residential mobility and to place it on a behavioral footing. However, the work in residential mobility is still relatively less concerned with cognitive processes than some other substantive fields within geography. In summary, it is the adoption of certain procedures and approaches (drawn from psychology) by geographers and others interested in residential mobility that has given research in that area a stronger behavioral component than would result from a simple concern with individual longitudinal files.

Perhaps the most significant criticism that can be leveled at 'behavioral geography' or the 'behavioral methodology' is that it has now served its purpose. It has moved geographic research from a concern with structure to a greater emphasis and concern with process and with the variables that underlie decision making in the spatial environment. It has emphasized not only process but also the importance of studying the individual elements that make up the process. Yet there are 'behavioral geographers' who would argue for the development of a general behavioral theory; and, more recently, activity-systems analysis has been proposed as an encompassing theory for behavioral investigations.[64] However, at the present time the results of activity-systems analysis (which have been going on in one form or another in planning, geography and sociology for the past dozen years)[65] have not yielded a consistent and coherent set of explanations for spatial behavior.

An even harsher critique of the behavioral approach is an extension of the earlier discussion of the trade-off between studies of individual decision making and analyses of supply-side impacts on residential-relocation behavior; at one extreme this has completely rejected the usefulness of behavioral analyses. Pahl suggests that we should 'concern ourselves with understanding the constraints and let the choices look after themselves,'[66] and Short argues that 'the rising tide of interest in residential mobility may be sweeping onto a deserted beach.'[67] On the other hand, it is clear that the emphasis on the supply side can include a discussion of behavior in the housing market, and that this perspective is central to studies focusing on the policy perspectives on residential mobility and the housing market.

Mobility, behavior and policy

The concern with policy implications has added another 'behavioral' dimension to research in residential mobility. Harris and Moore, in particular, have been concerned with assessing the larger impacts of mobility. It is their belief that the impacts of residential mobility are not just confined to individual and household behavior, but, rather, are experienced in an aggregate form by a whole range of social organizations. As a consequence, there must be more emphasis on residential mobility in context than on modeling individual behavior itself.[68] To examine residential mobility only in an individual-behavioral context would not provide information on the broader social context. To paraphrase their argument, it is the behavior of society and the mobility behavior within that larger behavioral matrix that are of interest.

There is already some evidence that recent studies of residential mobility have moved away from the earlier focus on attributes of moves and movers to a consideration of their role in the broader context of housing-market processes, neighborhood change and wider social contexts.[69] Several authors have discussed the ways in which household decisions are constrained and mediated by a variety of groups and institutions who operate in the housing market.[70] The dynamic relationship between individual choice and institutional behavior is increasingly emphasized. While there is an increasing ambivalence about purely 'behavioral' studies of residential mobility, it seems likely that the behavioral approaches and the attempts to build specific theory of individual residential mobility will not be abandoned. Rather, studies of mobility will be broadened to include analyses of individual behavior *within* the constraints of the wider structure, which was certainly one of the initial objectives formulated in the behavioral methodology.[71]

Conclusion

In this survey of residential mobility, we have observed a sequence of research interests that began with aggregate studies (population movement and redistribution), concentrated increasingly on individual movements and decision making, and finally focused on the links between behavior and structure. There is a parallel within the development of behavioral geography, which developed in response to the perceived inadequacies of studies of structure, and emphasized individual behavior and process. However, while studies in behavioral geography became increasingly intertwined with the models of psychology, there has been greater experimentation within the substantive field of residential mobility and a less specific concern with cognitive structures and mental processing. Moreover, the recent move to integrate behavior and structure, motivated by interest in the policy relevance of mobility research, is in advance of the present work in 'behavioral geography.' Only in consumer behavioral studies is there a similar thrust to link behavior and structure in meaningful terms.

Notes

1 A. Pred (1967) *Behavior and Location, I*, Lund, Gleerup; K. R. Cox and R. Golledge (1969) *Behavioral Problems in Geography: A Symposium*, Evanston, Ill., Northwestern University, Department of Geography, Studies in Geography 17.

2 P. H. Rossi (1955) *Why Families Move: A Study in the Social Psychology of Urban Residential Mobility*, New York, Free Press.

3 S. Goldstein (1954) 'Repeated migration as a factor in high mobility rates,' *American Sociological Review*, 19, 536–41; S. Goldstein (1958) *Patterns of Mobility, 1910–1950 (The Norristown Study)*, Philadelphia, University of Pennsylvania Press.

4 C. Myers, R. McGinnis and J. Pilger (1963) 'Internal migration as a stochastic process,' Ottawa, Paper given to the Thirty-fourth International Statistical Institute.

5 J. Wolpert (1965) 'Behavioral aspects of the decision to migrate,' *Papers of the Regional Science Association*, 15, 159–69; J. Wolpert (1967) 'Distance and directional bias in inter-metropolitan migration streams,' *Annals of the Association of American Geographers*, 57, 605–16.

6 R. G. Golledge (1970) *Process Approaches to the Analysis of Human Spatial Behavior*, Ohio State University, Department of Geography, Discussion Paper.

7 D. Harvey, 'Conceptual and measurement problems in geography,' in Cox and Golledge, *Behavioral Problems in Geography*, 35–68 (see note 1 above); G. Olsson, 'Inference problems in locational analysis,' in Cox and Golledge, ibid., 14–34.

8 G. Rushton (1969) 'Analysis of spatial behavior by revealed space preferences,' *Annals of the Association of American Geographers*, 59, 391–400.

9 H. Simon (1957) *Models of Man*, New York, Wiley.

10 P. Burnett (1976) 'Behavioral geography and the philosophy of mind,' in R. Golledge and G. Rushton (eds) *Spatial Choice and Spatial Behavior*, Columbus, Ohio State University Press.

11 ibid.

12 L. J. King and R. G. Golledge (1978) *Cities, Space and Behavior*, Englewood Cliffs, N.J., Prentice-Hall.

13 The literature on residential mobility has increased to the point where there are now a number of extensive and general reviews. J. M. Quigley and D. H. Weinberg (1977) 'Intraurban residential mobility: a review and synthesis,' *International Regional Science Review*, 2, 41–66, surveys current empirical research and emphasizes the lack of coherent theory. E. G. Moore (1972) *Residential Mobility in the City*, Washington, D.C., Association of American Geographers, Resource Paper 13, emphasizes geographic and spatial research on mobility. A. Speare, S. Goldstein and W. H. Frey (1975) *Residential Mobility, Migration and Metropolitan Change*, Cambridge, Mass., Ballinger, covers the literature in demography, sociology and geography. R. P.

Shaw (1975) *Migration Theory and Fact*, Philadelphia, Regional Research Institute Bibliography Series 5, is an annotated bibliography, while W. A. V. Clark and K. L. Avery (1978) 'Patterns of migration: a macro-analytic case study,' in D. Herbert and R. J. Johnston (eds) *Geography in the Urban Environment*, New York, Wiley, surveys the aggregate analyses of population movements within cities. An earlier review essay, J. W. Simmons (1968) 'Changing residence in the city: a review of intraurban mobility,' *Geographical Review*, 58, 622–51, analyses studies of residential mobility through the 1940s and 1950s.

14 Clark and Avery, 'Patterns of migration' (see note 13 above); W. Tobler (1978) 'Migration fields,' in W. A. V. Clark and E. G. Moore (eds) *Population Mobility and Residential Change*, Evanston, Ill., Northwestern University Press, 215–32.

15 Rossi, *Why Families Move* (see note 2 above); Goldstein, *Patterns of Mobility* (see note 3 above); Wolpert, 'Behavioral aspects of the decision to migrate' (see note 5 above); R. McGinnis (1968) 'A stochastic model of social mobility,' *American Sociological Review*, 23, 712–22. In reviewing the seminal literature that has influenced residential mobility I have chosen to cite McGinnis (1968) as a summary work because it reflects the earlier studies (Myers, McGinnis and Pilger 'Internal migration as a stochastic process' (see note 4 above); and G. Myers, R. McGinnis and G. Masnick (1967) 'The duration of residence approach to a dynamic stochastic model of internal migration: a test of the axiom of cumulative inertia,' *Eugenics Quarterly*, 14, 121–6), and is a good summary statement of the axiom of cumulative inertia.

16 Goldstein, 'Repeated migration as a factor in high mobility rates' (see note 3 above); Goldstein, *Patterns of Mobility* (see note 3 above).

17 Myers, McGinnis and Pilger, 'Internal migration as a stochastic process' (see note 4 above); and Wolpert, 'Behavioral aspects of the decision to migrate' (see note 5 above).

18 Wolpert, ibid., 160.

19 Myers, McGinnis and Pilger, 'Internal migration as a stochastic process' (see note 4 above); and McGinnis, 'A stochastic model of social mobility' (see note 15 above).

20 For example, E. W. Butler, F. S. Chapin, Jr, G. C. Hemmens, E. J. Kaiser, M. A. Stegman and S. F. Weiss (1969) *Moving Behavior and Residential Choice: A National Survey*, Washington, D.C., Highway Research Board, National Cooperative Highway Research Program Report 81. Much of this work has been codified in Speare, Goldstein and Frey, *Residential Mobility, Migration and Metropolitan Change* (see note 13 above).

21 R. Ginsberg (1971) 'Semi-Markov processes and mobility,' *Journal of Mathematical Sociology*, 2, 233–62; R. Ginsberg (1972) 'Critique of probabilistic models: application of the semi-Markov model to migration,' *Journal of Mathematical Sociology*, 2, 63–82; R. Ginsberg (1973) 'Stochastic models of residential and geographic mobility for

heterogeneous populations,' *Environment and Planning*, 5, 113–24; R. Boudon (1973) *Mathematical Structures of Social Mobility*, San Francisco, Ca., Jossey-Bass.

22 Wolpert, 'Behavioral aspects of the decision to migrate' (see note 5 above).

23 ibid.; Speare, Goldstein and Frey, *Residential Mobility, Migration and Metropolitan Change* (see note 13 above); J. S. Adams (1969) 'Directional bias in intraurban migration,' *Economic Geography*, 45, 302–23; L. A. Brown and E. Moore (1970) 'The intraurban migration process: a perspective,' *Geografiska Annaler*, B, 52, 1–13.

24 E. W. Butler, G. Sabagh and M. van Arsdol (1964) 'Demographic and social psychological factors in residential mobility,' *Sociology and Social Research*, 48, 139–54.

25 Quigley and Weinberg, 'Intraurban residential mobility,' 49 (see note 13 above).

26 E. A. Hanushek and J. M. Quigley (1978) 'An explicit model of intrametropolitan mobility,' *Land Economics*, 54, 411–29; E. A. Hanushek and J. M. Quigley (1978) 'Housing market disequilibrium and residential mobility,' in Clark and Moore (eds) *Population Mobility and Residential Change*, 51–98 (see note 14 above).

27 Quigley and Weinberg, 'Intraurban residential mobility,' 56 (see note 13 above).

28 W. Alonso (1964) *Location and Land Use*, Cambridge, Mass., Harvard University Press; R. Muth (1969) *Cities and Housing*, Chicago University Press.

29 Quigley and Weinberg, 'Intraurban residential mobility' (see note 13 above).

30 S. Spilerman (1972) 'The analysis of mobility processes by the introduction of independent variables into a Markov chain,' *American Sociological Review*, 37, 277–94; S. Spilerman (1972) 'Extensions of the mover–stayer model,' *American Journal of Sociology*, 78, 599–626.

31 Ginsberg, 'Semi-Markov processes and mobility'; 'Critique of probabilistic models'; 'Stochastic models of residential and geographic mobility for heterogeneous populations' (see note 21 above).

32 For a summary see, R. J. Pryor (1976) 'Conceptualizing migration behavior: a problem in microdemographic analysis,' in L. A. Kozinski and J. W. Webb (eds) *Population at Micro Scale*, New Zealand Geographic Society, Commission on Population Geography, 105–19.

33 Brown and Moore, 'The intraurban migration process' (see note 23 above).

34 L. A. Brown and D. B. Longbrake (1970) 'Migration flows in intraurban space: place utility considerations,' *Annals of the Association of American Geographers*, 60, 368–84.

35 W. A. V. Clark and M. Cadwallader (1973) 'Locational stress and residential mobility,' *Environment and Behavior*, 5, 29–41.

36 Speare, Goldstein and Fry, *Residential Mobility, Migration and Metropolitan Change* (see note 13 above).

37 J. O. Huff and W. A. V. Clark (1978) 'Cumulative stress and cumulative inertia: a behavioral model of the decision to move,' *Environment and Planning*, A, 10, 1101–9.

38 P. R. Gould and R. White (1974) *Mental Maps*, Harmondsworth, Penguin Books.

39 M. J. Boyle and M. E. Robinson (1979) 'Cognitive mapping and understanding,' in D. T. Herbert and R. J. Johnston (eds) *Geography in the Urban Environment: Progress in Research and Applications*, New York, Wiley, 59–82.

40 T. R. Smith (1978) 'Uncertainty, diversification and mental maps in spatial choice problems,' *Geographical Analysis*, 10, 120–41.

41 Adams, 'Directional bias in intraurban migration' (see note 23 above).

42 W. A. V. Clark (1971) 'A test of directional bias in residential mobility,' in H. McConnell and D. Yaseen (eds) *Models of Spatial Variation*, DeKalb, Ill., Northern Illinois University Press; B. Donaldson (1973) 'An empirical investigation into the concept of sectoral bias and the mental map: search, spaces and migration patterns of intraurban migrants,' *Geografiska Annaler*, 55, 13–33; R. J. Johnston (1971) 'Mental maps of the city: suburban preference patterns,' *Environment and Planning*, 3, 63–72.

43 Rushton, 'Analysis of spatial behavior by revealed space preferences' (see note 8 above).

44 W. A. V. Clark (forthcoming) 'A revealed preference analysis of intraurban migration choices.'

45 E. S. Shepherd (1979) 'Notes on spatial interaction,' *The Professional Geographer*, 31, 815.

46 J. R. Short (1977) 'The intraurban migration process: comments and empirical findings,' *Tijdschrift voor Economische en Sociale Geografie*, 68, 362–71; J. R. Short (1978) 'Residential mobility,' *Progress in Human Geography*, 2, 419–47; K. Bassett and J. R. Short (1980) *Housing and Residential Structure*, London, Routledge & Kegan Paul.

47 C. A. Maher (1980) 'Intraurban mobility and urban spatial structure: a review of current issues,' Melbourne, Australia, Monash University, Department of Geography, unpublished paper.

48 R. G. Golledge (1977) 'Behavioral approaches in geography: content and prospects,' in R. Taaffe and J. Odland (eds) *Geographical Horizons*, Dubuque, Iowa, Kendall Hunt, 87–96.

49 P. R. Gould, 'Cultivating the garden: a commentary and critique on some multidimensional speculations,' in Golledge and Rushton, *Spatial Choice and Spatial Behavior* (see note 10 above).

50 J. Piccolo and J. Louviere (1977) 'Information integration theory applied to real-world choice behavior: validational experiments involving shopping and residential choice,' *Great Plains Rocky Mountain Geographical Journal*, 6, 49–63.

51 T. R. Smith, W. A. V. Clark, J. O. Huff and P. Shapiro (1979) 'A decision-making and search model for intraurban migration,' *Geographical Analysis*, 11, 1–22.

52 W. A. V. Clark and T. R. Smith (1979) 'Modeling information use in

a spatial context,' *Annals of the Association of American Geographers*, 69, 575–88.

53 See, for example, P. R. Gould (1975) 'Acquiring spatial information,' *Economic Geography*, 51, 87–99; R. G. Golledge and G. Moore (1976) *Environmental Knowing: Theories, Research and Methods*, Stroudsburg, Pa., Dowden, Hutchinson & Ross; M. Cadwallader (in press) 'Urban information and preference surfaces: their patterns, structures and interrelationships,' *Geografiska Annaler*.

54 See, for example, T. Hägerstrand (1967) *Innovation Diffusion as a Spatial Process*, University of Chicago Press; L. A. Brown (1968) *Diffusion Processes and Location: A Conceptual Framework and Bibliography*, Philadelphia, Regional Science Association Institute.

55 R. G. Golledge and L. A. Brown (1967) 'Search, learning and the market decision process,' *Geografiska Annaler*, 49, 116–24.

56 See, for example, P. Burnett (1973) 'The dimensions of alternatives in spatial choice problems,' *Geographical Analysis*, 5, 181–204; S. R. Lieber, 'A comparison of metric and nonmetric scaling models in preference research,' in Golledge and Rushton, *Spatial Choice and Spatial Behavior* (see note 10 above).

57 J. J. Louviere and R. J. Meyer (1976) 'A model of residential impression formation,' *Geographical Analysis*, 8, 479–86; J. J. Louviere and D. Henley (1977) 'Information integration theory applied to student apartment selection decisions,' *Geographical Analysis*, 9, 130–41.

58 Huff and Clark, 'Cumulative stress and cumulative inertia: a behavioral model of the decision to move' (see note 37 above); Speare, Goldstein and Frey, *Residential Mobility, Migration and Metropolitan Change* (see note 13 above).

59 ibid.; Hanushek and Quigley, 'An explicit model of intrametropolitan mobility'; 'Housing market disequilibrium and residential mobility' (see note 26 above); Huff and Clarke, 'Cumulative stress and cumulative inertia' (see note 37 above).

60 Smith, Clark, Huff and Shapiro, 'A decision-making and search model for intraurban migration,' 7 (see note 51 above).

61 D. Weinberg *et al.* (1977) *Housing Allowance Demand Experiment. Locational Choice, Part I: Search and Mobility*, Cambridge, ABT Associates.

62 Speare, Goldstein and Frey, *Residential Mobility, Migration and Metropolitan Change* (see note 13 above).

63 More general descriptive studies of search behavior have been undertaken in F. A. Barrett (1973) *Residential Search Behavior*, Toronto, York University, Geographical Monographs, and in R. Flowerdew (1976) 'Search strategies and stopping rules in residential mobility,' *Transactions of the Institute of British Geographers*, 1, 47–57, although their main contribution has been to suggest hypotheses for formal modeling and testing.

64 A. Pred (1977) 'The choreography of existence: comments on Hägerstrand's time geography and its usefulness,' *Economic Geography*, 53, 207–21.

65 I. Cullen and V. Godson (1972) *The Structure of Activity Patterns: A Bibliography*, London University, Joint Unit for Planning Research, Research Paper 2.
66 R. Pahl (ed.) (1975) *Whose City?* Harmondsworth, Penguin Books, 156.
67 Short, 'Residential mobility,' 421 (see note 46 above).
68 E. G. Moore and R. S. Harris (1979) 'Residential mobility and public policy,' *Geographical Analysis*, 11, 175–83; R. S. Harris and E. G. Moore (1980) 'An historical approach to mobility research,' *Professional Geographer*, 32, 22–9.
69 W. A. V. Clark and E. G. Moore (eds) (1980) *Residential Mobility and Public Policy*, Beverly Hills, Ca., Sage Publications.
70 For an overview, see E. G. Moore and W. A. V. Clark, 'The policy context for mobility research,' in Clark and Moore, ibid., and, for a specific example, M. Dear, 'The public city,' in Clark and Moore, ibid.
71 Cox and Golledge, *Behavioral Problems in Geography* (see note 1 above).

PART 3

9 Behavioral geography and the philosophies of meaning

David Ley

It is perhaps an indication of human geography's growing maturity as a social science that after its long period of empirical tranquility and reluctance to engage methodological, let alone philosophical, issues, the subject has attempted over the past fifteen years to assimilate quantification, behavioralism and, more recently, humanism and structural marxism. This condensation of intellectual history, which in other disciplines was spread over a period of two or three times as long, has created a turbulence which, while exciting, can also be confusing, as the intellectual half-life of not only theories but also whole paradigms shrinks rapidly. In true North American fashion, obsolescence is setting in more and more speedily; North American geography, too, has its annual ritual of spring cleaning and its Easter parade of new models each April, when last year's motifs are cleared from the intellectual wardrobe. More reflective and synthetic work is clearly needed to evaluate and trace the course of distinctive research traditions, to show the nature of continuity and interrelation between them, and to place narrow debate within its broader theoretical and historical context.

This is a broad mandate, and one to which the present chapter will make only a modest contribution. This chapter is concerned to outline the relations between the geographic schools of behavioralism and humanism, which represent one manifestation of the broader relations between the philosophy of logical positivism and what I will call the philosophies of meaning, including such traditions as phenomenology, existentialism and pragmatism. These positions, we will see, differ substantially on the question of the ontological status and methodological examination of subjectivity. In the first part of the paper the humanistic critique of behavioralism will be presented; this will be followed by a review of humanistic work itself; finally, some of the challenges raised against geographic humanism will be discussed, leading to a conclusion that, after the schismatic recent history of human geography, the time is now approaching for the abandonment of narrow dogmatic claims to universalism by any particular position, in favor of efforts at integration leading to a higher order of synthesis.

The critique of behavioralism

During the 1960s a significant theoretical break occurred in human geography with the development of perception studies emphasizing the variable and contingent nature of the environment. The theoretical transition from environment to behavior and thus to landscape could not be accomplished by holding constant human values and initiative. Physical resources were not simply objective givens of the physical environment; they were objects endowed with meaning by a geographic subject, a fusion of objectivity and subjectivity. This realization occurred in two separate literatures. Cultural and historical geographers showed, even for simple societies, the important role of human judgement in resource evaluation and the development of place. This tradition was illustrated in Brookfield's seminal essay 'On the environment as perceived,' which laid out an interpretative human geography in which contextual methods like participant observation and ethnoscience would uncover the typical meanings and actions that gave character to a place.[1] Mercer and Powell saw in such a cognitive–behavioral approach a contribution that would 'preserve and foster a "humanist" alternative to the popular mechanistic explanation.'[2] So, too, it was this more catholic definition which encouraged my own participant observation study in inner-city Philadelphia to claim an affinity with behavioral geography.[3]

But it was an alternative view of behavioralism that prevailed, set out in a second review paper by Downs, which proposed that behavioral geography could be the science of spatial behavior and spatial decision making, a science committed to prediction and statistical explanation, rather than to contextual understanding.[4] In the disciplinary setting of the quantitative revolution and the ascendancy of positivist philosophy, the assimilation of behavioralism was inevitable. The cognitive–behavioral approach was subsumed by spatial analysis as 'an appendage' to the locational school.[5] Within this conceptualization, the full force of the scientific method was applied in, for example, a preoccupation with measurement, operational definitions and a highly formalized methodology. Subjectivity, in short, was to be confined within the straitjacket of logical positivism.

It is worth noting that it was in precisely such an intellectual milieu that phenomenology was born as, initially, a critical philosophy. Husserl, the founder of phenomenology, countered the application of positivist method to the human sciences in his day:

> In the second half of the nineteenth century the world-view of modern man was determined by the positive sciences. That meant a turning away from the questions which are decisive for a genuine humanity . . . . Mere factual knowledge makes for factual men . . . . In our desperate need, this science has nothing to say to us.[6]

Later, in Watson's stimulus–response behaviorism, subjectivity was not merely repressed, but was altogether rejected. Watson dismissed the role of human intelligence, and regarded the behavior of men, like that of animals, to be propelled by a conditioned reflex to environmental stimuli. Such a conceptualization was forcefully disapproved by American pragmatists, notably G. H. Mead, and in Europe by Alfred Schutz, the phenomenologist who has had perhaps the greatest influence on contemporary social science. 'It is not then quite understandable,' Schutz mused, 'why an intelligent individual should write books for others or even meet others in congresses where it is reciprocally proved that the intelligence of the other is a questionable fact.'[7] Gestalt psychology represented a withdrawal from the extreme claims of behaviorism in as much as perception was granted a central theoretical role. But nevertheless, claimed the French phenomenologist, Merleau-Ponty, it had not broken free of the epistemological limitations of psychophysics.[8] In Lewin's field theory, for example, motivation was regarded as a product of positive and negative field forces, a position reminiscent of behaviorism.[9] Even if such subjective traits as perception and motivation were now granted an existence, it was not a separate existence, for subjectivity was still seen as an outcome of determining environmental factors. The view of man remained essentially passive. As we will see, these criticisms are of more than historical interest.

The differences between behaviorism and behavioralism are not as substantial as they might seem. In sociology, 'Watsonian behaviourism might be dead, but it is surely resuscitated both in the letter and the spirit of the behavioural science label.'[10] So too a critic of Schutzian phenomenology has nonetheless observed that 'the differences between the psychological theory of behaviorism and the broader approach to social science called behavioralism are based on convenience, not principle.'[11] Both share a common philosophy of science that includes a pallid model of man. In political science, behavioralism has implied the full baggage of positivism with its aspirations to the methods of the natural sciences.[12] In sociology, behavioralism has ushered in strict objectivity, the experimental method, quantification and reductionism.[13]

It is scarcely surprising that a similar assessment could be made of behavioral geography, for it shares a common pedigree in behaviorism and gestalt psychology with the behavioralism practised in other social sciences. Behaviorism has been revived in B. F. Skinner's operant conditioning, and geographic theories of, for example, search and learning have made considerable use of reward and reinforcement as independent variables in formulations where learning is seen as a function of environmental states.[14] So, too, Harvey has commented that 'the necessary operational concepts for the measurement of human perceptual behavior [sic] are provided by stimulus–response psychology.'[15] Again, gestalt psychology has been a significant influence on behavioral geography. Kirk's pioneering discussion of the behavioral environment was drawn from Wertheimer, while Kurt Lewin has exercised a profound effect on decision-making studies in geography, where such important concepts as 'aspiration level' and 'gatekeeper' are derived from his social psychology.[16] Thus the criticisms against behaviorism and gestalt psychology by an earlier generation of phenomenologists and also against behavioralism in other social sciences are still current in an evaluation of contemporary behavioral geography.

These criticisms are commonly directed against the assumption of naturalism, the practice of reductionism, and the epistemology and methodology of positivism, characteristics that are closely related.

Naturalism sees no essential discontinuity between man and nature, so that a common scientific mode of explanation may be applied indiscriminately in the human as in the natural sciences. Such a viewpoint inevitably lays stress on environing forces controlling behavior, which is interpreted in terms of inevitable response or even conditioned reflex. Naturalism has had an influential role in human geography, including such venerable traditions as Ratzel's stern injunction that 'A people should live on the land fate has given them; they should die there submitting to the law.' The association between behavioral geography and animal behavior has revived an extreme naturalism, as such traits as exploration, spacing, crowding, territoriality and aggression have been related to etho-logical experiments. While only a few behavioral geographers have made this particular association, more widespread features of their work, in particular a general commitment to the methodological ideal of the physical sciences, identify them with a naturalist position.[17]

The absorption of man into naturalist explanation invariably implies a reduction of what it is to be human. A shrunken and essentially passive view of man is characteristic of naturalist explanation, including the concept of economic man, which re-mained an important normative model for behavioral geogra-phers.[18] Human consciousness is given little theoretical status, for a naturalist theory of perception ascribes primacy to external stimuli, the objects of perception which may be measured by scientific technique, while minimising the active (or 'distorting') effects of human cognition. Consciousness is thereby seen as a product of external factors, for it is part of naturalist doctrine that 'for every set of attributes (or variables) there is some system which is deter-ministic with respect to that set.'[19] A more extreme behaviorist position dispenses with consciousness altogether, dismissing the attributes of mind, such as choice, thought, emotion and will as 'the phlogiston of the social sciences,' to use George Lundberg's astonishing phrase.

In contrast, humanistic philosophies reject the extension of naturalism to the study of human action and consciousness, for the social sciences are regarded as qualitatively separate from the natural sciences. The distinction is provided by human intention-ality; in Merleau-Ponty's phrase, in social life we are condemned to meaning. This introduces an important differentiation:

The world of nature, as explored by the natural scientist, does not 'mean' anything to the molecules, atoms, and electrons therein. The observational field of the social scientist, however . . . has a specific meaning and relevance structure for the human beings living, acting, and thinking therein.[20]

Much, though not all, of the subject matter of the social scientist consists of 'facts' which are endowed with intentionality. Thus they are to some extent contingent and variable, rather than fixed and universal. The same wooded hillside, which may be analysed confidently and unambiguously by a botanist or geomorphologist, may become variable in the light of human perception when viewed by a forester, a property developer, or a representative of the Sierra Club. Meanings, moreover, are contextual, so that the same object may be viewed differently on separate occasions by a single observer as the context changes. The facts of human geography cannot be viewed independently of a subject whose concerns confer their meaning, a meaning that directs subsequent action. Unlike the natural sciences, then, the social sciences cannot escape the task of interpreting the domain of consciousness and subjectivity, a domain that precludes the granting of an invariant and universal status to objects or relationships.

 Some of the implications of this distinction are indicated by considering the difference between understanding and explanation. Positivist explanation proceeds by uncovering regularities existing between measurable observations: if A, then B. We might attach an explanatory level to the relationship in probabilistic terms, using correlation or some other statistic. But this is often a weak criterion for explanation; the demonstration of a sequence between events need not necessarily lead to an understanding of the linking process.[21] In an example we will return to later, the demonstration of an apparently invariant functional relationship between the objective distance to a town and the perceived distance, gives no understanding of *why* the relationship takes the form it does. The weakness here is a weakness of positivist philosophy 'that we cannot have knowledge of anything but observable phenomena and of the relations between them.'[22] This limitation is particularly crippling when extended to subjectivity, for subjective states are not observable and measurable phenomena in the same sense as objects are. As a result subjectivity raises severe methodological problems for positivist method. One response, as we have seen, is the behaviorist assertion that the categories of mind do not exist. Behavioralism usually falls short of so extreme a position but continues to be embarrassed in its treatment of subjectivity: 'the behaviourally oriented sociologist finds it difficult to speak in terms of values, attitudes, consciousness.'[23]

Since mental states are only approachable, according to the positivist, through their manifestations, various diversions have been employed in behavioral geography to force objectivity onto consciousness. One tactic has been to infer prior, unobservable mental states on the basis of consequent observable behavior, as in Rushton's analysis of revealed locational preferences that are assessed retroactively on the basis of actual spatial behavior.[24] To achieve more than the truism that people shop at stores they prefer, Rushton argues for the existence of a latent structure of preferences. This structure is exposed using the black box procedure of scaling. But what such analysis glosses over is that the use of scaling (or indeed other mathematical techniques like factor analysis) as a *heuristic*, together with the subsequent *interpretation* of frequently ambiguous dimensions (or factors), introduces a mystique, as a whole series of assumptions enter the analysis which are of an essentially *qualitative* nature. Such methodology acts almost to provide some *deus ex machina*, producing a structure where none is immediately observable, a practice at odds with the necessity for objectively observable phenomena required by positivist philosophy. Such an essentialist practice is, from a positivist stance, open to the charge of being metaphysical, or at the very least is open to subjectivism on the part of the researcher himself.

This statement may be illustrated from the literature of psychophysics, which, as its name suggests, aspires to a full naturalist epistemology with its objective of discovering universal laws of mental states. But from the application of psychophysics to behavioral geography, it is likely that these laws are attained in large part as an artifact of the researcher's *own* subjectivity. The literature on subjective or perceived distance estimates, for example, has the goal of discovering an invariant law linking objective phenomena and subjective states within the setting of a laboratory experiment. The subjects in the experiment are a student group preselected for their *homogeneity*, which is 'one of the preconditions for obtaining regular data.'[25] The stimuli are a set of cities, from which a smaller set are chosen because they are the most *similar* in terms of the subjects' knowledge, interest and estimated importance. Finally, in the analysis individual responses are *averaged out* in tracing the association between perceived and objective distance to the towns. The result is an apparently invariant relationship, which the researcher raises to the status of a law.

But it could be persuasively argued that such a regularity is little more than the inevitable fabrication of a contrived methodology. So much empirical variety has been defined away by the experimental method that the discovery of a regularity in the data is not only unsurprising, but also probably specious. This criticism may be extended to other areas of behavioral geography characterized by an overpreoccupation with methodology. The use of formalized instruments such as the semantic differential or the grid repertory test, with their multivariate statistical back-up to reveal hidden structures, glosses over a number of qualitative problems, not least the difficulty faced by the man in the street (as opposed to under-graduates) in completing them.[26] One suspects that the empirical data generated by such methods may often be little more than a polite fiction.

A trenchant criticism against what he called abstracted empiricism, with its surrender to methodology, was made by the pragmatist C. Wright Mills a decade before geographic behavioralism developed. He noted that:

> There is a pronounced tendency to confuse whatever is to be studied with the set of methods suggested for its study . . . their most cherished professional self-image is that of the natural scientist . . . . What they have done, in brief, is to embrace one philosophy of science which they now suppose to be The Scientific Method. The vehicle of their abdication is pretentious over-elaboration of 'method' and 'theory'; the main reason for it is their lack of firm connection with substantive problems.[27]

This lack of groundedness was an inevitable consequence of the law-seeking quest, for to establish theory and laws it was necessary that the 'accidents' of empirical reality be removed to lay bear the 'universals.' For the positivist project to succeed, phenomena and relations had to be abstracted, removed from their contexts. The laboratory and the student sample are the logical expression of behavioralism; here (but, one suspects, only here), in sterile environments and with subjects wrenched from their everyday worlds, the researcher's ability to manipulate stimuli may lead to the discovery of 'laws.' This may appear a harsh designation, but intrusive methodology, where the researcher explicitly manages the research situation while dismembering the contexts of spontaneous action, is a pervasive characteristic of behavioral geography. The removal of context in all its forms is perhaps the major weakness of behavioral method, for, while regularities may be claimed from the study of rarified man in simplified settings, there is no guarantee that this abstraction bears any relation at all with the concrete actions of everyday life, where it is precisely this contextuality, which positivism so carefully removes, that ascribes meaning to actions.

Such concepts as place utility, preference functions, and satisficing man (who is no more than the negation of optimizing rational man) remind us of behavioral geography's close association with locational analysis, which is itself a derivative of neoclassical economics. It is therefore no surprise that the subjects of behavioralism, whether they are encountered in the classroom, in the parking lot of a supermarket, or on their own doorsteps, are treated as isolated individuals. They are systematically detached from the social contexts of their actions. And yet, as humanistically oriented work in social geography has indicated, it is precisely within the contexts of a social world that actions originate and have their meaning.[28] An understanding of such actions cannot proceed in separation from the social milieus to which they are dialectically bound.

But how can such milieus be penetrated and known? The answer lies in what we might call methodologies of engagement, methodologies that enable the researcher to understand and interpret the subjective-meaning contexts of individuals and groups in their own milieu, in as non-obtrusive a manner as possible. This requires detailed familiarity and immersion within the contexts of action. My own research in the black inner city of Philadelphia made use of participant observation, non-obtrusive indicators, literary, artistic and folk sources, as well as institutional records and the minutes of voluntary organizations. At the conclusion of the research period these approaches were supported by the more obtrusive method of a field questionnaire. For historical research, the data will inevitably be less rich, so that in one sense we can have less confidence that we have successfully penetrated the subjective meaning contexts of our predecessors.[29] However, from another perspective, the temporal distance from another society may be more enriching for the researcher, for the alternative frame of reference may throw into doubt, and therefore make self-conscious, his own taken-for-granted presuppositions.[30]

The interpretative methodology of *verstehen* (or, more currently, hermeneutics) points to a divide between the natural and social sciences that throws into doubt the naturalist assumption at the root of behavioralism. Whereas the positivist method of the former is directed toward explanation through the detection of regularities, the interpretative method of the latter is directed toward understanding in the recovery of meanings.[31] This is not to claim a complete break between the sciences, for there are problems in the social sciences where positivist method is appropriate (for example, the description of commodity flows), and indeed some authors would find a place for interpretative understanding in the natural sciences.[32] But, in contrast to the universal claims of geographic positivism in the 1960s, there is now wide agreement among social theorists from varying backgrounds of the necessity for an interpretative epistemology of social life.[33] Indeed, in contemporary social theory it is positivism that is very much on the defensive.

Geographic humanism

In claiming an identity for itself, geographic humanism has played a significant role in the critique of positivism. But the critique also has a deeper motivation, for phenomenologists in particular have always maintained that in its foundational, taken-for-granted conventions, positivist methodology does in fact provide a stunning example of the primacy of an unreflected world of shared meanings in the construction of both facts and theory. Moreover, it was inevitable that the early programmatic papers advocating an anthropocentric perspective in geography derived from the philosophies of meaning should counter the prevailing positivist orthodoxy. But, as we will see, this does not imply, as some recent critiques of humanism have suggested, that the perspective has only a critical role to play in the development of theory in geography.[34]

Since the programmatic and conceptual arguments for geographic humanism have now been made, they will not be repeated here.[35] Rather, it might be more useful to review some of the substantive contributions that have followed broadly humanistic currents. Inevitably much of this work is recent or still in progress.

The body of research most commonly associated with the humanist venture is what we might call the sense of place studies; it was, for example, the *only* substantive contribution discussed by Entrikin in his critique of geographic humanism. This work, associated particularly with the literary writings of Yi-Fu Tuan and David Lowenthal, seeks to identify the dominant meanings of a place and the quality of geographical experience.[36] In Relph's analysis of the meaning of the suburban landscape and Seamon's reflections on the everyday experience of place, an explicitly phenomenological underpinning is claimed.[37] The elegant, perceptive and scholarly nature of much of this writing has often been noted, and yet I do not feel it should be seen as more than one possible model for humanistic research, or at least for humanistic research, which would aspire also to the status of social science. For its method is essentially that of focused intuition, separate often both from empirical method and from a specific empirical problem. The meanings identified are not necessarily grounded in empirical settings; they do not engage a concrete problem. Or if, as with Relph's study of placelessness, an existential problem is identified, there is a methodological subjectivity in approaching it; Relph's discussion of an authentic landscape, for example, depends heavily upon a set of personal aesthetic judgements.

A separate body of more grounded empirical work has traced the relationship between landscape and identity. It has followed in large part the tradition of G. H. Mead's symbolic interactionism, itself a development of pragmatist philosophy, and an important school of empirical sociology in North America. The central argument is that place is a negotiated reality, a social construction by a purposeful set of actors. But the relationship is mutual, for places in turn develop and reinforce the identity of the social group that claims them. In an early paper, Duncan discussed how, in a high-status village within commuting range of New York, a twofold division within the upper middle class could be related not only to segregation patterns and membership in voluntary associations, but also to the decoration and landscaping of the home and surrounding lot. The dichotomy was sharpened in a study of traditional and modern elites in an Indian city, where the selection of a home in the old town or a modern suburb presented an explicit message of identity and intimated a whole set of life–world relations.[38] The home is the most articulate landscape expression of the self and can reinforce either a positive self-image, or, in the case of dreary public housing in an unwanted location, it may sustain an identity of a peripheral and low-status member of society with little ability to mold his environment.[39]

There is, then, a reciprocal symbolic interchange between people and places. Other research has probed the geographical experience of such constrained groups as the tramp, the urban Indian, and the elderly. Rowles' study of the elderly was also of methodological interest as it proceeded beyond participant observation to the method of interpersonal knowing.[40] The author developed close friendships over time with just five elderly people, maintaining that the richness of detail and understanding gained by such immersion within the lifeworlds of others offset the limitations of so small a sample size. In each of these research endeavors, the aim has been to reconstitute the subjective meanings of individuals and groups, in order to understand their actions and the meanings that places hold for them. There has been an attempt to gain an insider's view, the definition of the situation by individuals in the constitution of their social worlds and their experience of place. This literature has had policy implications in the recognition of the multiple realities of urban life and the need to incorporate more actively the views of clients as well as managers in social policy. Programs of decentralization and greater public participation spring from the necessity to plan in a manner that acknowledges the importance of frames of reference that may well depart from those of policy makers. In an innovative piece of phenomenologically informed research, Godkin has shown how places signify a whole set of personal meanings and interactions that may be used to advantage in therapeutic counselling.[41] The reciprocal status of place and identity suggested the use of 'place chronologies' as a focus for clinical interviews and treatment, and this idea has been adopted in a successful experiment treating alcoholics.

There are other implications to the topographies of meaning ascribed to geographic space. Like other commodities, space is engaged not only as a brute fact, but also as a product with symbolic meaning. Space offers more than its own dimensionality; as place it is a vicarious commodity pointing beyond its physical or functional nature to a set of values which may be appropriated. In post-industrial society consumption does not entail the simple acquisition of an object but also the appropriation of a nonmaterial or symbolic value. Consumers are interested in what places will do for them. So it is that the value of land and consumer preferences are no longer assessed according to conventional measures of accessibility alone, but also according to such existential gradients as status, security, stimulation, danger and amenity.[42] During the 1960s, for example, the five fastest growing metropolitan regions in the United States were all recreation or retirement centers. In post-industrial society, increasingly it is the *meaning* of space that is setting the price gradient and directing population movements.

Places may also be examined more deeply to reveal less self-conscious or explicit messages of dominant ideologies and social relations. From a close examination of the townscape of an early working-class suburb in Vancouver, Gibson has claimed to identify a set of foundational, if not always self-conscious, beliefs held by the first settlers.[43] This treatment of landscape as a product that may cast light on the values and perceptions of historical groups is a procedure that has been much used by cultural and historical geographers. But landscape cannot always be taken at its face value, for an interpretation of underlying values should go beyond the viewpoint of the historic actor to consider the social relations out of which values themselves emerge. This is not to reinstate a new determinism but rather to emphasize the dialectical nature of human freedom. While men create landscapes, they do so within constraints that may not always be visible to them. Even the great men of history make decisions in context; though Chairman Mao represented a fresh start in the molding of the Chinese landscape he was, if in part unknowingly, at least as much a victim of past Chinese tradition as he was the creator of a new one.[44]

This dialectic relationship between creativity and constraint holds in all empirical settings. Evans has shown that the suburbs are not, as some have claimed, simply the product of a technological or economic necessity, but are also the active creation of their residents.[45] So too, even in the constrained world of the black inner city, there is no functional necessity for juvenile delinquency to occur unless there is an active interpretation of objective constraints into a subjectively defined situation, an interpretation that often does not happen.[46] Such a conjunction between structural constraints and subjective meanings need not occur, so that both objectivity and subjectivity, both causes and reasons, and both explanation and understanding need to be a part of the research methodology.

There are three further points concerning a theory of action in human geography. Firstly, it may be extended beyond informal groups to include formal associations.[47] Secondly, relations *between* groups should be examined as well as relations internal to a social lifeworld; such study will inevitably deal with the distribution of power among different interests. Thirdly, just as the precedents for an action may in part be unrecognized, so too their consequences may not always be as expected. Lowman has suggested in his examination of the settings of street prostitution that an active interpretation of the law by the police may well have real effects (spatial and otherwise), though these effects may be counter-intuitive in terms of police expectations.[48] So too the routine decisions of a whole series of urban managers, be they public-housing officials, bank managers or social workers, will have real consequences in the urban experience of clients, even if these consequences are unintended. In an ambitious attempt to examine the interacting lifeworlds of British Columbia Indians and the federal Department of Indian Affairs, Kariya is showing that the everyday experience of Indians is in part a product of the typifications of the federal managers.[49] While the structural factors can rarely be disregarded, nonetheless, the subcultural values of managers have a significant and often autonomous effect at the level of experience, a truth acknowledged in their own empirical analysis even by critics of a managerial thesis. Even explicitly structural analysis thereby shows as much support for Weber's theory of bureaucracy as for Marx's thesis of class conflict; as one such study concluded, 'access to housing is bureaucratically defined.'[50] What could be more Weberian than that?

Some remaining challenges

There is, then, good empirical support for the proposition that a human geography building from the philosophies of meaning may be empirically and theoretically fruitful, and be able to make a contribution that is not merely critical or negative of existing orthodoxy. Nevertheless, several shortcomings have been noted in a humanistic perspective, and these will be discussed briefly. I will suggest that each of these criticisms holds only for a limited and overly idealistic form of humanism.

The most common criticism directed against Mead's interaction-ism or Schutzian phenomenology is that it neglects the role of structures and power relations which constrain an individual's freedom and can produce the potential for conflict between groups. At one level this argument is well directed for both schools emphasize in particular the internal culture-building relations of a social world. But there are also ample evidences of external constraints in, for example, Schutz's writing. He speaks of the inherited biographical situation of individuals, such objective factors as age, sex, race, and time and space, which bound a person's actions; he develops also the reactive 'because of' motives, which are set against active 'in order to' motives in an interpretation of an individual's intentions. 'We are, however, not only centers of spontaneity, gearing into the world and creating changes within it, but also the mere passive recipients of events beyond our control which occur without our interference.'[51] While it is true that Schutz's method does not generally treat an individual's biographical situation as problematic, yet his anthropo-centric theory of action is capable of extension to analyze such factors explicitly. For example, in their influential treatise on the sociology of knowledge Berger and Luckmann use Schutz's phenomenology of the lifeworld as the first phase in a three-phase model that includes institutional factors.[52] Again, as Silverman's discussion suggests, in the examination of institutions there is a theoretical affinity between Schutzian phenomenology and Weberian institutional analysis.[53] Moreover, Schutz's concept of multiple realities admits of dissimilar and potentially conflicting perceptions. Finally, and countering the criticism that humanist man is granted too much theoretical freedom, a recent critique of Schutzian phenomenology has concluded that the lifeworld exercises too tight a control on individual thought and action.[54]

Related to the neglect of structural realities is the assertion that humanistic research satisfies itself with unique or trivial subject matter, that it is an expression of naive empiricism, or, presumably worse, bourgeois sentimentality. A selective argument could present supporting evidence, most notably in the tedious literature of some ethnomethodology with its painstaking description of mundane and theoretically insignificant interactions.[55] But phenomenology is by no means committed to such unedifying description of the self-evident. Husserl's project was to uncover the *hidden* structures of the lifeworld and, although his methodology of transcendental reductions is now little more than a curiosity, the commitment remains to a foundational criticism that makes questionable the unquestioned and throws into profile the taken for granted. Neither was Husserl concerned with intellectual minutiae, for his famous *Krisis* lectures, in which he developed his theory of the lifeworld, were motivated in large part by the growing barbarism he detected in European society in the 1930s, a barbarism he claimed that was not unrelated to the success of a naturalism in science that rejected human subjectivity. So, too, it has been humanistically inclined writers who have been centrally concerned with such existential themes as authenticity and self-actualization, dehumanization and alienation in contemporary society.

Aligned to the criticism of trivial and unique subject matter has been the charge that humanistic perspectives do not engage in the scientific parade of objectivity, classification, and the development of theory and, ultimately, laws. As it stands, this argument is one that is made entirely from within a positivist perspective, and as such its presuppositions would be unacceptable to social scientists who do not subscribe to a naturalist epistemology. The charge of methodological subjectivity has been answered many times. As Schutz noted, the method of *verstehen* or interpretative understanding does not require introspection or an unverifiable empathy, but is based upon the same principles of intersubjectivity, the sharing of meanings, that we exercise successfully in our everyday lives as joint members of society.[56] In one striking illustration he notes how in a court of law the attempt by the jurors to reconstitute the motives of the accused require anything but uncontrollable introspection or empathy.

Moreover the methodology of *verstehen* (and its redefinition in hermeneutics) may lead to generalization as the researcher establishes 'objective meaning-contexts of subjective meaning-contexts.' The vehicle here has often been the ideal type. In everyday life, actors establish generalizations or typifications of the meanings and intentions of others that have a twofold adequacy. They have both a *meaning adequacy*, that is they represent satisfactorily the subjectivity of the other, and also a *causal adequacy* so that they anticipate successfully a predictable stream of behavior. So too in social science the task is to construct ideal types that satisfactorily meet these criteria of adequacy. For a social-science ideal type to be valid it must pass the test both of meaning and causal adequacy. This conceptualization of the ideal type has generated considerable controversy.[57] But its heuristic value is abundantly demonstrated by Weber's own use of the ideal type, notably in *The Protestant Ethic and The Spirit of Capitalism*, which illustrates also the commitment of humanistically oriented research to generalizable concepts, even if it would not subscribe to the positivist aspiration to develop laws along the lines of the natural sciences.

A third major complaint against humanistic research is its pre-occupation with process rather than with effects, which is seen as a symptom of the larger problem of an overassociation with idealistic forms of thought. A fixation upon consciousness and the forms of perception is said to eclipse both the material preconditions and the material consequences of thought and action. Again there is some basis for such a criticism, for in geographic inquiry both J. K. Wright's geosophy and the more recent sense of place studies are characterized by a thinly veiled idealism where meanings or perceptions often seem to be separate from a material world of preconditions and consequences. But such a restricted basis is not necessary in humanistic research, and the more recent position of phenomenology and existentialism is that social life must not be compromised by the study of pure consciousness; inverting Husserl, the orthodox position is now that man's essence is his existence.[58]

These three criticisms of geographic humanism are inter-connected, for a charge of idealism is invariably associated with the complaint of an overactive view of man that neglects structural and political constraints and leads to a contemplation of unique histories and geographies. I have suggested here that on theoretical and philo-sophical grounds these criticisms may be circumvented and that humanism is not confined to the limiting possibilities of Husserlian, or any other, idealism. But the more convincing rebuttal comes from the substantive literature of geographic humanism itself, where, as we saw earlier, the commitment to existence rather than essence is accompanied by a discussion of material and power relations and the examination of contexts that include, but also fall beyond, the consciousness of man, the geographic agent.

Conclusion

In this chapter I have argued that the absorption of geographic behavioralism into the severe epistemological mold of positivism has prompted the development of humanistic approaches to deal with questions of subjectivity from the more appropriate epistemological source of the philosophies of meaning, which include such compatible traditions as existentialism, phenomenology and pragmatism. Among a number of important thinkers in these schools who might qualify for inclusion, I have emphasized the contribution of Alfred Schutz, whom one critic has called 'the single most important and influential figure in the movement to phenomenologically re-define, or "humanize," social science.'[59] Schutz's project was to bind Husserl's philosophy of the lifeworld with Weber's theory of social action, but critics, especially in human geography, have perhaps made too much of the connection with Husserl and too little of the connection with Weber. Human geography has been overly influenced by first Durkheim and more recently Marx, especially those elements of their thinking closest to naturalism and positivism. Weber's commitment to an interpretative and contextual social science without exclusive alignment to either materialism or idealism might well exemplify a useful theoretical corrective.

In any case, it is certain that human geography will have to take the realms of subjectivity and the lifeworld more seriously than behavioralism has allowed. There is a growing consensus in the social sciences, both in review volumes and in original works, that adequate explanation of human actions must include both the fatalism of social structures and the creative spontaneity of the lifeworld. To cite just one example from many in contemporary social theory: 'the primary problem is how to develop theories that satisfactorily synthesize the structural analysis of different social formations, and the explanation of human action in terms of subjective states and meanings.'[60] Or, as C. Wright Mills put it, the task is to develop a theoretical position that brings together both public issues and private troubles. This is an important challenge to human geography, which in its theoretical development has progressively lost contact with man, the geographic agent. As Schutz has intimated, there may be a high price for such theoretical abstraction: 'The safeguarding of the subjective point of view is the only but sufficient guarantee that the world of social reality will not be replaced by a fictitional non-existing world constructed by the scientific observer.'[61]

Notes

1 H. Brookfield (1969) 'On the environment as perceived,' *Progress in Geography*, 1, 51–80.
2 D. Mercer and J. Powell (1972) *Phenomenology and Related Non-Positivistic Viewpoints in the Social Sciences*, Melbourne, Australia, University of Monash Publications in Geography 1.
3 D. Ley (1974) *The Black Inner City as Frontier Outpost: Images and Behavior of a Philadelphia Neighborhood*. Washington, D.C., Association of American Geographers, Monograph Series 7.
4 R. Downs (1970) 'Geographic space perception,' *Progress in Geography*, 2, 67–108.
5 D. Harvey (1969) 'Conceptual and measurement problems in the cognitive–behavioral approach to location theory,' in K. Cox and R. Golledge (eds) *Behavioral Problems in Geography: A Symposium*, Evanston, Ill., Northwestern University Press, Studies in Geography 17, 35–67.
6 Quoted in F. Buytendijk (1967) 'Husserl's phenomenology and its significance for contemporary psychology,' in N. Lawrence and D. O'Connor (eds) *Readings in Existential Phenomenology*, Englewood Cliffs, N.J., Prentice-Hall, 352–64.
7 A. Schutz (1970) *On Phenomenology and Social Relations* (ed. H. Wagner), University of Chicago Press, 266.
8 M. Merleau-Ponty (1964) *The Primacy of Perception*, Evanston, Ill., Northwestern University Press.
9 K. Lewin (1935) *A Dynamic Theory of Personality*, New York, McGraw-Hill.
10 A. Brittan (1973) *Meanings and Situations*, London, Routledge & Kegan Paul.
11 R. Gorman (1977) *The Dual Vision*, London, Routledge & Kegan Paul.
12 For a generally critical discussion of behavioralism in political science and other social sciences, see M. Natanson (ed.) (1973) *Phenomenology and the Social Sciences*, Evanston, Ill., Northwestern University Press.
13 Brittan, *Meanings and Situations* (see note 10 above).
14 R. Golledge, 'The geographical relevance of some learning models,' in Cox and Golledge, *Behavioral Problems in Geography*, 101–45 (see note 5 above).
15 Harvey, 'Conceptual and measurement problems' (see note 5 above).
16 W. Kirk (1973) 'Problems of geography,' *Geography*, 48, 357–71; R. Downs and D. Stea (eds) (1973) *Image and Environment*, Chicago, Aldine, 5–6.

17 The theme of the methodological unity of the sciences around the ideal of the physical sciences is argued in D. Harvey (1969) *Explanation in Geography*, London, Arnold, and in D. Amedeo and R. Golledge (1975) *An Introduction to Scientific Reasoning in Geography*, New York, Wiley. The latter book is significant for its treatment of behavioral geography (especially Chapter 12).

18 Harvey, for example, found it to be 'an extraordinarily fruitful' concept (Harvey, 'Conceptual and measurement problems' (see note 5 above)). Compare the severe criticisms offered in M. Hollis and E. Nell (1975) *Rational Economic Man*, Cambridge University Press.

19 E. Nagel (1961) *The Structure of Science: Problems in the Logic of Scientific Explanation*, New York, Harcourt, Brace & World, 595. Cited in Gorman, *The Dual Vision* (see note 11 above).

20 A. Schutz (1954) 'Concept and theory formation in the social sciences,' *Journal of Philosophy*, 51, 257–73.

21 R. Keat and J. Urry (1975) *Social Theory as Science*, London, Routledge & Kegan Paul.

22 ibid., 70.

23 Brittan, *Meanings and Situations*, 20 (see note 10 above).

24 G. Rushton, 'The scaling of locational preferences,' in Cox and Golledge, *Behavioral Problems in Geography*, 197–227 (see note 5 above).

25 S. Dornic (1967) 'Subjective distance and emotional involvement: a verification of the exponent invariance,' Reports from the Psychological Laboratories, University of Stockholm, 237.

26 See, for example, the discussion in R. Downs (1970) 'The cognitive structure of an urban shopping center,' *Environment and Behavior*, 2, 13–39.

27 C. Wright Mills (1959) *The Sociological Imagination*, Oxford University Press, Chapter 3.

28 For an example see, D. Ley (1975) 'The street gang in its milieu,' in G. Gappert and H. Rose (eds) *The Social Economy of Cities*, Beverly Hills, Ca., Sage, 247–73. For a theoretical discussion, see D. Ley (1978) 'Social geography and social action,' in D. Ley and M. Samuels (eds) *Humanistic Geography*, Chicago, Maaroufa Press, 41–57.

29 E. Gibson, 'Understanding the subjective meaning of places,' in Ley and Samuels, ibid., 138–54; C. Harris, 'The historical mind and the practice of geography,' in Ley and Samuels, ibid., 123–37.

30 This argument is raised in a geographical context by Gregory, who notes that Schutz's method with its incomplete discussion of historical actions cannot offer such reflexivity, or self-awareness, to the researcher. However, this is not quite true, for, though Schutz paid less attention to historical distance, he did examine cultural distance, and how a meeting between culturally distant individuals can jolt and make problematic everyday taken-for-granted attitudes and actions. D. Gregory (1978) 'The discourse of the past: phenomenology, structuralism, and historical geography,' *Journal of Historical Geography*, 4, 161–73; A. Schutz (1944) 'The stranger,' *American Journal of Sociology*, 49, 499–507.

31 For a discussion of hermeneutic method in geography following Dilthey, Weber and Schutz, see C. Rose (1977) 'The concept of reach and its relevance to social geography,' Clark University, unpublished dissertation.

32 Keat and Urry, *Social Theory as Science*, 174–75 (see note 21 above).

33 For example, see ibid.; A. Giddens (1977) *New Rules of Sociological Method*, London, Hutchinson; J. Habermas (1971) *Knowledge and Human Interests*, Boston, Beacon Press. In geography this position is present in G. Olsson (1975) *Birds in Egg*, University of Michigan, Geographical Publications 15; D. Gregory (1978) *Ideology, Science and Human Geography*, London, Hutchinson; Ley and Samuels, *Humanistic Geography* (see note 28 above).

34 J. Entrikin (1976) 'Contemporary humanism in geography,' *Annals of the Association of American Geographers*, 66, 615–32.

35 E. Relph (1970) 'An inquiry into the relations between phenomenology and geography,' *Canadian Geographer*, 14, 193–201; Y-F. Tuan (1971) 'Geography, phenomenology and the study of human nature,' *Canadian Geographer*, 15, 181–92; Mercer and Powell, *Phenomenology and Related Non-Positivistic Viewpoints in the Social Sciences* (see note 2 above); A. Buttimer (1976) 'Grasping the dynamism of life-world,' *Annals of the Association of American Geographers*, 66, 277–92; D. Ley (1977) 'Social geography and the taken-for-granted world,' *Transactions of the Institute of British Geographers*, NS, 2, 498–512.

36 For representative statements, see Y-F. Tuan (1974) *Topophilia*, Englewood Cliffs, N.J., Prentice-Hall; Y-F. Tuan (1977) *Space and Place: The Perspective of Experience*, Minneapolis, University of Minnesota Press; D. Lowenthal and H. Prince (1965) 'English landscape tastes,' *Geographical Review*, 55, 186–222.

37 E. Relph (1976) *Place and Placelessness*, London, Pion; D. Seamon (1979) *A Geography of the Lifeworld*, London, Croom Helm.

38 J. Duncan (1973) 'Landscape taste as a symbol of group identity,' *Geographic Review*, 63, 334–55; J. Duncan and N. Duncan (1976) 'Social worlds, status passage and environmental perspectives,' in G. Moore and R. Golledge (eds) *Environmental Knowing*, Stroudsburg, Pa., Dowden, Hutchinson & Ross.

39 C. Cooper (1974) 'The House as a symbol of the self,' in J. Lang *et al.* (eds) *Designing for Human Behavior*, Stroudsburg, Pa., Dowden, Hutchinson & Ross; D. Sopher (1979) 'The landscape of home,' in D. Meinig (ed.) *The Cultural Meaning of Ordinary Landscapes*, New York, Oxford University Press.

40 G. Rowles (1978) *Prisoners of Space?* Boulder, Col., Westview Press.

41 M. Godkin (1977) 'Space, time and place in the human experience of stress,' Clark University, unpublished dissertation.

42 D. Ley (forthcoming) *A Social Geography of the City*, New York, Harper & Row.

43 Gibson, 'Understanding the subjective meaning of places' (see note 29 above).

44 M. Samuels, 'Individual and landscape: thoughts on China and the Tao

of Mao,' in Ley and Samuels, *Humanistic Geography*, 283–96 (see note 28 above).

45　D. Evans (1978) 'Demystifying suburban landscapes,' Portland, Paper presented to the Association of Pacific Coast Geographers.

46　Ley, 'The street gang in its milieu' (see note 28 above).

47　D. Silverman (1970) *The Theory of Organizations*, London, Heinemann.

48　J. Lowman (1979) 'The geography of crime: a critique of the analytical separation of crime and justice,' Victoria, Paper presented to the Canadian Association of Geographers.

49　P. Kariya (1978) 'Keepers and kept: the lifeworld relations of British Columbia Indians and the Department of Indian Affairs,' New Orleans, Paper presented to the Association of American Geographers.

50　J. Lambert, C. Paris, B. Blackaby (1978) *Housing Policy and the State*, London, Macmillan, 150.

51　A. Schutz (1946) 'The well-informed citizen: an essay on the social distribution of knowledge,' *Social Research*, 13, 463–78.

52　P. Berger and T. Luckmann (1966) *The Social Construction of Reality*, Garden City, N.Y., Doubleday.

53　Silverman, *The Theory of Organizations* (see note 47 above).

54　Gorman, *The Dual Vision* (see note 11 above).

55　But see also an interesting defence of ethnomethodology in Keat and Urry, *Social Theory as Science*, 174 (see note 21 above).

56　Schutz, 'Concept and theory formation in the social sciences' (see note 20 above).

57　For recent illustrations of this debate, compare the critical exposition of Gorman, *The Dual Vision* (see note 11 above) with the favourable discussion in R. Williame (1973) *Les Fondements Phénoménologiques de la Sociologie Compréhensive: Alfred Schutz et Max Weber*, The Hague, Nijhoff.

58　For expositions of this orthodoxy see, for example, W. Luijpen and H. Koren (1969) *A First Introduction to Existential Phenomenology*, Pittsburgh, Duquesne University Press; M. Roche (1973) *Phenomenology, Language, and the Social Sciences*, London, Routledge & Kegan Paul.

59　Gorman, *The Dual Vision*, 2 (see note 11 above).

60　Keat and Urry, *Social Theory as Science*, 229 (see note 21 above).

61　Schutz, *On Phenomenology and Social Relations*, 271 (see note 7 above).

10 Of paths and projects: individual behavior and its societal context

Allan Pred

A poet spends his life in repeated projects, over and over again trying to build or dream the world in which he lives. But more and more he realizes that this world is at once his and everybody's. It cannot be purely private, any more than it can be purely public. It grows out of a common participation which is nonetheless recorded in authentically personal images.

Thomas Merton, *The Geography of Lograire*

Most common individuals, the people you and I interact with every day on a fleeting or more lasting basis, differ from the poet only in their degree of awareness of, and sensitivity to, the surrounding world. Bard or not, each and every person in his private life is affected by the public world, or society, around him. Yet each and every person is in some way a part of the public world, and thereby affects the private life content, or external actions and internal experiences, of others. In order to gain insight into individual behavior, it is

This chapter was written while the author was receiving research support from the US National Science Foundation. Many of the ideas and arguments it contains have been further developed by the author since its completion in early 1979. See A. Pred (1981) 'Production, family and "free time" projects: a time-geographic perspective on the individual and societal change in nineteenth-century US cities,' *Journal of Historical Geography*, 7; A. Pred (1981) 'Social reproduction and the time geography of everyday life,' *Geografiska Annaler*, B, 63; A. Pred (1981) 'Power, everyday practice and the discipline of human geography,' in A. Pred, *Space and Time in Geography: Essays Dedicated to Torsten Hägerstrand*, Lund Studies In Geography, B, 47; and A. Pred and N. Thrift (1981) 'Time geography: a new beginning,' *Progress in Human Geography*, 5(2).

necessary to hold insights as to the workings of society. In order to gain insights into the workings of society, it is necessary to possess some insights into individual behavior. Especially in a world where human transactions have become increasingly fragmented and complicated due to the high division of labor and extreme functional specialization that characterizes contemporary large-scale organizations and institutions; especially in a world where alienation and a sense of powerlessness are increasingly common elements in the daily experiences of people, it would seem appropriate – even urgent – that so-called 'behavioral' and 'modern human' geographers try to understand individual behavior in terms of its societal context, or in terms of the dialectical relationships between the individual and society.[1]

Despite their supposed concern with either the 'spatial organization of *society*,' or some aspect of 'man–land' relationships (wherein '*man*' must have a collective or societal dimension as well as an individual dimension), 'behavioral' and 'modern human' geographers have shown little proclivity to deal with the societal context of individual behavior. Instead, in keeping with the infiltration of division-of-labor and specialization principles into academia in general, 'behavioral' and 'modern human' geographers have usually been content to approach individual behavior in a reductionist and piecemeal fashion that has stressed 'space (or distance), measurability, and visual landscape . . . [or] the surface features of the external.' And: 'Since the external is in things rather than relations, we have produced studies of reifications in which man, woman, and child inevitably are treated as things and not as the sensitive, constantly evolving human beings we are.'[2] The journey to work and to shop, the recreational trip, the migratory move, the perception of some portion of the environment, the entrepreneurial locational decision, and the reaction to some natural hazard, are each normally analyzed at the individual level as self-contained behaviors that are totally divorced and unrelated to any past actions or experiences of the observed people-things, and that are completely uninfluenced by the particular workings of institutions and society in general. Or behavior outcomes are aggregated in a gravity model or some other timeless quantitative expression in a manner that frequently obliterates the interdependence of actions taken by different people, the delicate and intricate connections between past and present behaviors of the same person, and the influence of opportunities and constraints, as well as encouragements and discouragements embedded within the societal context.

For a variety of reasons, the yet-emerging conceptual framework and philosophical perspective of time geography lends itself very well to the task of trying to understand individual behavior in terms of its societal context. By virtue of its providing a common language to cover different kinds of interactions both between humans and between individuals and elements of the man-made and natural environment, by dealing with 'the fundamental bonds between life, time and space,'[3] by being plastic enough to deal with the actions of specific persons as well as the temporal and spatial organization of entire activity systems, the time-geographic approach definitionally provides an easily crossed two-way bridge between microlevel (individual life content) and macrolevel (societal) scales of observation. Time geography is also well suited as a point of departure for analyzing what goes on at the interface between individual behavior and the workings of society, since it is a holistic approach which, by explicitly recognizing the physical indivisibility of each person, does not remove the actions and role fulfilments of individuals from their sequential and locational context. It therefore allows otherwise seemingly unrelated activities, events and collateral processes that involve institutions, technology and the natural environment to be seen as expressions of the complexly interwoven 'local connectedness' associated with society's temporal and spatial organization. Time geography is also an appealing means by which initially to confront questions concerning the societal context of individual behavior since it allows one to be at the same time humanistic in outlook and deductive in reasoning – a combination of circumstances that often allows considerable insights to be extracted from comparatively scanty information.[4]

Although several seminal works remain inaccessible to those lacking a command of Swedish, it is not the intent of this article to present a comprehensive review of the time-geography framework and the literature associated with it; for several published works, available in English, already perform that service in part, if not in whole.[5] Instead, attention will be focused on two concepts – paths and projects – that are central to the time-geographic approach to individual behavior and its societal context. Not only will the basic attributes of paths and projects be considered, but those concepts

will also be related to the dialectical relationships that are entrenched deeply in the time-geographic perspective on the human
condition. In addition, brief consideration will be given to two
somewhat similar overviews of individual behavior and its societal
context, each of which is intimately related to time geography in
general, and paths and projects in particular.

Some basic attributes of paths and projects

Paths

At the heart of Hägerstrand's time geography, as initially formulated, is the notion that all of the actions and events that sequen-
tially make up the individual's existence have both temporal and
spatial attributes, and that consequently the biography of a person
is always on the move with her and can be depicted diagra-
matically at daily, weekly, annual or life-long scales of observation
as an unbroken, continuous path through time–space. (In
Hägerstrand's framework, all living organisms belonging to the
animal and vegetable worlds and all man-made objects including
buildings, are also thought of as having 'biographies' and describing
continuous paths in time–space.) The simultaneous weaving out of
paths by all the human individuals (and tools, machinery, and other
inanimate objects) in a given area can thus be thought of as a
constantly unfolding and complex web. Under such circumstances,
the most noteworthy events occur at specific stations – labeled
'stations' or 'domains' – either when two or more paths are coupled
or grouped together to form an 'activity bundle,' or when such
groups are broken up, with one or more of the participants moving
or in space and time sooner or later to become associated with other
'activity bundles' at other 'stations.' Each activity bundle that an
individual finds her way to is not regarded as something that is to be
classified and isolated with all other activities of a similar type joined
by the person in question, but instead is viewed in its sequential and
interdependent context. Consequently, a focus on paths to some
extent becomes synonymous with a concern for the ways in which
time-geographic 'realities' and constraints circumscribe the
individual's ability to move from station to station and choose
among activity bundles.

The events and actions incorporated into an individual's path are limited, most basically by: her indivisibility, or inability to participate simultaneously in spatially separated activities; the limited duration of her own life, as well as the limited duration of the lives of other people and inanimate objects; the finite time resources at her command each day, and the inescapable facts that all tasks have a duration and that all movement between spatially separated points is time consuming; her restricted ability to partake in more than one task at a time; and the fact that space has a limited packing capacity or ability to accommodate events (because no two physical objects can occupy the same space at the same time).

The action and event sequence, or behavioral choices that accumulate along an individual's path on a *day-by-day* basis, also may be thought of as being hemmed in by three major types of constraints:

> *Capability constraints* circumscribe activity participation by demanding that large chunks of time be allocated to physiological necessities (sleeping, eating and personal care) and by limiting the distance an individual can cover within a given timespan in accord with the transportation technology [at her command]. *Coupling constraints* pinpoint where, when and for how long the individual must join other individuals (or objects) in order to form production, consumption, social and miscellaneous activity bundles. *Authority constraints* . . . subsume those general rules, laws, economic barriers and power relationships which determine who does or does not have access to specific domains at specific times to do specific things.[6]

At lengthier scales of observation, the composition of an individual's path will be circumscribed by the number and mix of specialized, independently existing, roles proffered up by the organizations and institutions found within any given bounded area (or by the 'activity system' of a given area), and by the rules, competency requirements, and economic, class and other constraints that govern entrance to those roles. (The specialized roles embedded within organizations and institutions exist independently – until terminated or superseded – in the sense that when they are not filled by one person, they sooner or later must be filled by another person.)

Projects

The path perspective, and its attendant emphasis on behavioral restrictions, can easily lead to the mistaken conclusion that time geography, insofar as it is concerned with individual–societal relationships, is merely a form of constraint analysis that is best suited for exercises in social engineering. Such a mistaken conclusion, which is probably one of the key reasons why the time-geographic framework has not gained more widespread acceptance, can be dispelled at least partially by a consideration of the less familiar project concept and some of its implications.

From the time-geographic viewpoint, a project consists of *the entire series of tasks necessary to the completion of any goal-orientated behavior*. Whether behavior stems from the establishment of goals by individuals, groups, or coalitions ranging 'from the family upward in size to the state, the multinational corporation and international organizations,'[7] the project concept is equally applicable. A project may consist of things as mundane as the making of a bed, the planting of a garden flower, or the writing of a letter. Or, a project can involve a chain of tasks as complicated as that necessary for organizing and carrying out a professional meeting, for drafting and gaining legislative approval of a new law, or for producing a finished good. The tasks associated with a simple or a complex project usually have an internal logic of their own which requires that they be sequenced in a more or less specific order. (In a bed-making project, sheets and blankets must be straightened out before and not after placing the cover; in a professional-meeting project, paper sessions must be organized before and not after the opening ceremonies.)

Each of the logically sequenced component tasks in a project is synonymous with the formation of activity bundles, or with the convergence in time and space of the paths being traced out either by two or more humans, or by one or more people and one or more physically tangible inputs or resources (e.g. equipment, raw materials, buildings).[8] Hence, *when an individual's path becomes wrapped up with the activity bundle(s) of a project defined by an organization or by some institution other than the family, there is a direct intersection between the external actions of the individual and the observable workings of society.*

Whether a project has a one-time or repetitive (routinized)

character, the paths that must be synchronized and 'synchorized' will not normally be identical for every constituent activity bundle. Some paths of necessity will be tied up with all or most of a project's activity bundles. Other paths often will have a more abbreviated association with a project (Figure 10.1). Moreover, the completion of more complex projects may require the carrying out of several subprojects that are temporally parallel to one another, and which definitionally cannot absorb the same paths for the period of time when they are simultaneous. During their development, more complex projects also may give rise to subordinate, unanticipated off-shoot projects with activity-bundle sequences of their own.[9]

Since project participation, by definition, places a demand on the limited daily time resources of the individual, since the daily time resources of groups and entire populations are limited to twenty-four hours times their number, and since projects must be carried out at one or more locations in space, any time–space region has a bounded capacity for accommodating the activity bundles spawned by projects. During the period of time when the paths of human beings (or other living organisms or man-made objects) are committed to one activity bundle, they cannot simultaneously join an activity bundle belonging to another project. Or project participation makes individuals temporarily inaccessible to other individuals who might wish to join them for the purpose of doing other things at the same place or at other locations. If, as a consequence, these individuals opt to form their own project-related activity bundle at another time, they in turn will become inaccessible to other persons at that new time, possibly pushing aside or completely eliminating the chances that yet other project-related activity bundles will ever materialize.

It is through such spread effects in conjunction with the carrying out of projects that a 'local connectedness' is created between otherwise seemingly unrelated activities, events and processes. Put otherwise:

> The whole operation of a society – [the projects connected with] economic production consumption, social interaction, [and] recreation . . . [as well as acts of] contemplation [carried out in personal isolation] – can be viewed as a huge competition between events and states for exclusive spaces and times.[10]

Thus, whatever the scale of a project, whatever its degree of

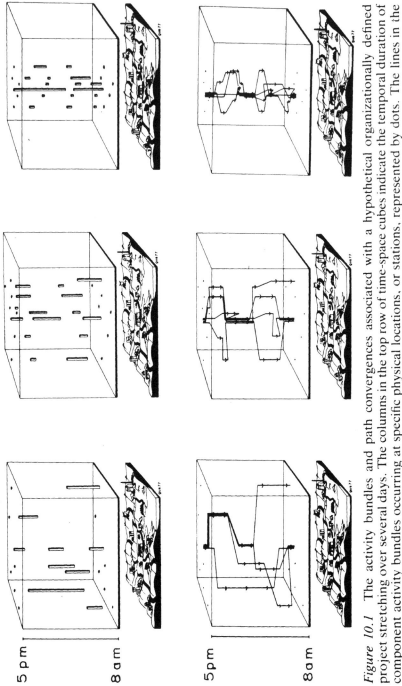

Figure 10.1 The activity bundles and path convergences associated with a hypothetical organizationally defined project stretching over several days. The columns in the top row of time-space cubes indicate the temporal duration of component activity bundles occurring at specific physical locations, or stations, represented by dots. The lines in the lower row of time-space cubes indicate the individual daily paths moving in and out of those same activity bundles. Derived from Olander 'Foretagsadministration som aktivitetssystem i tid och rum,' p. 61; and Cederlund, 'Administrativ verksamhet som projekt,' p. 79. (See note 9 for complete references.)

complexity, it cannot make its entrance, or become assimilated, into the time–space of a specified area unless: there already is a niche, or a series of openings, in the existing way of doing things, which will allow individuals (and any necessary natural or man-made objects) to couple their paths at required times and places to form activity bundles for required durations; or a niche is chiseled out to fit its time and space demands and path-participation requirements through the elimination, modification or rescheduling of previously routinized projects and their component activity bundles.[11]

Two very important differences often exist between projects that are the result of goals formulated, on the one hand, by single individuals or families (or households) and, on the other hand, by an organization or institution other than the family. (Organizations and institutions are nothing more than intellectual abstractions unless there are one or more physically observable projects with which they can be identified.) One of these differences relates to role delegatability and the other to the fixedness of activity bundles in time–space. When a project is based upon individual or family goals, the roles necessary to project completion are either self- or family assigned, and therefore may either, in the former case, be temporarily or permanently foregone (the live-alone bed maker may choose to leave her bed in an unkempt state), or, in the latter case, temporarily or permanently delegated to another family member (the family bed maker may request someone else to undertake her tasks). However, when a project is based upon the goals of an organization or nonfamily institution, and when component activity bundles require the presence of individuals holding independently existing roles within the organization or institution, those individuals – unless in positions of authority – are not normally free either to forego or delegate their role requirements.[12]

Similarly, when a project is based upon individual or family goals, the participant(s) in theory will have something to say, however small, about when (and perhaps where) activity bundles are to occur – especially since there are apt to be several self- or family-defined projects to juggle about (bed making, if it is to occur at all, may take place before or after the projects of dressing and breakfast preparation). But, when a project springs from the goals of an organization or nonfamily institution, any holder of an independently existing role within the organization or institution who becomes a participant will usually have to join one or more activity bundle at physically *fixed* locations. This may occur either during precisely *fixed* temporal periods whose duration equals the necessary task time(s), or during more flexible yet still *fixed* temporal periods whose duration somewhat exceeds the necessary task time(s).

The just-named differences should not obscure the fact that, especially under modern circumstances, where work is separated from home, *just when and where, if at all, an individual finds it possible to become part of the activity bundles belonging to self- or family-defined projects is in great measure determined by the fixed temporal and spatial attributes of the projects set in motion by organizations and outside institutions, or by the observable workings of society.* That is, owing to the operation of time-geographic 'realities,' once an individual's path becomes tied up with the projects and fixed activity bundles of an organization or institution associated with, for example, production, consumption, education or religion, her ability to schedule and participate in self- or family-defined projects becomes at least moderately constrained. And, once an individual must synchronize and 'synchorize' her path with the activity bundles of several organizational or institutional projects existing in parallel – as is generally necessary in today's highly specialized and organizationally dominated post-industrial societies – her ability to schedule and participate in self- or family-defined projects is apt to become highly constrained. Frequently high stress levels are created thereby, which frustrate the fulfilment of her emotional and psychological needs (as well as those of her husband or any other adult household companion she may have), and affect the socialization, personality development and sexual identity of her children (if she has any).

Moreover, as the individual in modern 'post-industrial' societies has 'to comply with a growing number of fixed time(–space) points' in order to participate in ever more finely divided organizational and institutional projects, she cannot take part in projects 'for maintaining, strengthening or renewing' her social network 'as simply or naturally as . . . [she] used to.'[13] This circumstance often compounds the sense of alienation she has acquired from accumulated acts of participation in activity bundles of the former projects, i.e. from spending so much of her time either in filling roles with an independent existence of their own, or in making fragmented, fleeting and highly impersonal contacts with other people who are themselves filling 'thingified' independently existing roles at the moment of encounter.

Paths, projects and the dialectics of time geography

The preceding discussion presumably has made it clear that the time-geographic concept of the project and its component activity bundles can be used as a starting focal point from which to examine the interface between individual behavior and the workings of society. However, there is a danger that the concept still leaves a lingering and mistaken sense that time geography is nothing but a form of constraint analysis. The richness of time geography in general, and the path and project concepts in particular, as a means for getting at individual–societal relationships, can be much better appreciated by turning to the four interrelated and overlapping dialectics that lie firmly beneath the time-geographic perspective on the human condition.

The life path–daily path dialectic

Nothing can touch an individual's life without influencing the course of her daily path, and nothing can influence the course of her daily path without having the potential to touch upon her life path. When an individual's life path becomes linked up with an independently existing role within an organization or institution, she must as a consequence repeatedly channel her daily path to activity bundles belonging to certain projects. Yet, as a result of participating in those activity bundles at precise geographic locations and more or less fixed temporal locations, and as a result of having her participation in other types of projects and activity bundles thus constrained, the individual is exposed daily to other people, inanimate objects, first- and second-hand ideas, and informational impulses in general that she otherwise would not have encountered. These exposures, in turn, help shape the choices she subsequently makes about which other long-term independently existing roles, if any, to seek. This is not only because it is in wandering over daily paths that individuals intentionally or unintentionally learn of the life-path opportunities society has to offer, but also because it is out of the soil of accumulated daily interactions – and reactions – that long-range intentions and choice motivations grow (see *The external–internal dialectic* below).

Different stages of what sociologists and others label the 'life cycle' also can be viewed in terms of the life path–daily path dialectic. When, at birth, a child's path becomes associated with a particular family and the place(s) at which that family dwells (or when a person is in the 'dependent child' stage of her life cycle), she is socialized in a given way, she learns a given language or dialect and its attendant ways of classifying and understanding information and experience, and her personality uniquely develops owing to her daily path encounters with her parents, siblings and relatives (if any), schoolmates and friends. In turn, these encounters – themselves very much the product of the way in which the life path–daily path dialectic has led her parents to specific independently existing roles with specific incomes, status and class attributes – are synonymous with the inculcation of values, attitudes, skills and preferences, which are eventually translated into intentions concerning the type(s) of independently existing roles, if any, to seek in organizations or nonfamily institutions after the 'dependent child' stage comes to an end.[15] Furthermore, if a person enters the various married-life stages of the life cycle, or when the *life paths* of two adults become parallel to one another in terms of residence, it is ultimately due (except in the case of a marriage contract) to an initial *daily path* interaction – an interaction that more likely than not was brought about directly or indirectly by the activity-bundle requirements of an organizational or institutional project.

The external (corporeal action)*–internal* (mental experience and intention) *dialectic*

The external and internal ought not be seen as dichotomous opposites, but as inseparable parts of the same dialectical whole: the life content of the individual. External behavior and internal 'experience are simultaneous and exist as an integrated and continuous flow in time and space.'[16] *On the one hand, no human individual can spin out her physically observable daily and life paths, corporeally (or externally) participating in projects, activity bundles and interactions with other persons and objects, without amassing mental (or internal) experiences that are fundamental to the shaping of her values, perceptions, attitudes, capabilities, preferences, and conscious or subconscious motivations, and hence her goals and intentions* (only some of which will be realized). On the other hand, *in choosing among projects* (or in choosing among roles, activity bundles and interactions that are possible within the environment's time-geographic constraints), *and in thereby eventually adding corporeal actions to her daily and life paths, no individual can escape the influence of her previous mental experiences and consequently derived goals and intentions.* [17] Or, as a person incessantly glides from past to future along the tip of a constantly advancing 'now' line, she is at the center of a repeated dialectical interplay between her corporeal actions and her mental experiences and intentions.

The operation of the external–internal dialectic allows individual behavior and its societal context to be viewed in at least two very important ways. Firstly, the following more specific restatement of the dialectic can be made. When parts of a person's daily and life paths are corporeally steered through time–space as a result of involvement in an organizational or institutional project, or as a result of involvement in the workings of society, she is confronted by environmental impulses, personal contacts, influences and information in general, as well as emotions and feelings, that she otherwise would not have experienced mentally. When a person consciously or unconsciously employs the mental experiences acquired through participation in organizational or institutional projects to formulate her goals and intentions, and chooses among alternatively possible projects, actions and interactions, she sooner or later undertakes corporeal action that otherwise would not have become a part of her daily and life paths.

A second view emerges if it is acknowledged that organizational and institutional projects and roles are not the outcome of spontaneous combustion or autonomous forces, but are instead the consequence of the goals established and decisions reached by individuals and coalitions who hold power and authority within organizations and institutions. For, if this is recognized then it also is to be acknowledged that the *internally* based inputs used in the formulation of those goals and decisions – whether singly or collectively arrived at – can be traced back to the previous *external* path and projection-participation history of the goal setters and decision makers themselves. Or, the project-defining goals and decisions arrived at by administrators and other organizational and institutional elites cannot be separated from the images of the environment they have built up from their limited acquisition of imperfect information, from any anticipation they may hold rewards or penalties, and from their absorption (or rejection) of cultural and ideological values and norms, all of which, in turn, cannot be separated from the course of their earlier daily and life paths.[18]

In a somewhat different but not totally unrelated vein, it is to be further realized that, even when a person chooses to reflect or contemplate for a given stretch of seconds, minutes or hours, the external–internal dialectic is in some sense at work. Insofar as reflection and contemplation are influenced by language, personally and culturally based symbols, memories and meanings ascribed to events and environmental stimuli, they are dependent upon the corporeal participation in projects, activity bundles and inter-actions that enable both the learning of language and symbols and the development of memories and meanings. Moreover, reflection or contemplation can only come about after a decision has been made – in however a dimly conscious manner – to bring one's body (or path) into conjunction with a given chair, bed or piece of ground. And reflection or contemplation cannot terminate without a decision to undertake a new activity (corporeal action) unless an individual drifts off to sleep.

The path convergence–path divergence or creation–destruction dialectic

Because of the corporeal nature of humans and their inability to defy gravity, the time–space trajectory of every individual is always proximate and contiguous to other people and objects, or is in physical touch and 'togetherness' with the time–space trajectories of other people and elements of the natural and man-made environ-ment. As a consequence, *no activity bundle or interaction can come into being via the convergence of paths without necessitating the divergence of previously joined paths; and no activity bundle or interaction can terminate, via the divergence of paths, without necessitating the convergence of paths in new combinations, or new activity bundles or interactions.* This dialectic is even in evidence at detailed scales of observation; for no person can take the smallest step without destroying one union between foot and ground, or floor, and creating another such union. It means, in other words, that everything is part of an unbroken chain of creation and destruction, everything is in perpetual transition, and 'we are all, human beings as well as stones, irresistibly part of the flow of emergence and disappearance.' Or, each 'here and now is the waist of a sandglass where biographies (or paths) first concentrate (in time–space) and then spread out.'[19] And, because of the nature of the path convergence–path divergence dialectic (operating alone or in combination with the life path–daily path and external–internal dialectics), every situation is inevitably rooted in past situations and inevitably becomes the root of subsequent situations.

The path convergence–path divergence dialectic has a clear expression at the interface between individual behavior and the workings of society. Each activity bundle included within an organizational or institutional project only becomes possible upon the dissolution of all those immediately prior activity bundles and also upon the dissolution of interactions involving project bundles' human participants and resource inputs. Conversely, no activity bundle associated with an organizational or institutional project can come to an end without all the participating individuals (as well as any surviving objects or newly created outputs) moving on in time–space to become, by choice or by necessity, part of immediately subsequent activity bundles or interactions, or path combinations, the composition of which is constrained by the time-geographic characteristics of the project bundle itself and the surrounding environment. In this light, any organizational or institutional project can be thought of alternatively as a loosely or rigidly choreographed sequence of created and destroyed states of individual 'togetherness.'

The path convergence–path divergence dialectic infuses new meaning into tilled fields, roads, buildings, furniture, machinery and other elements of the man-made environment with which an individual's path comes into touch. These objects are not lifeless, just as elements of the natural environment are not lifeless. They are, instead, to be seen as the outcome of numerous previous path convergences and path divergences involving people, other man-made objects, natural phenomena, or the outcome of projects – projects usually defined by societal organizations or institutions, projects that are synonymous with the transformation of nature. Man-made objects are also to be seen destined for destruction, or withering away, normally as a result of future path convergences and path divergences, or projects.[20]

The individual–societal dialectic

There is nothing that can affect the time geography, or path, of an individual without affecting the time-geographic workings of society as a whole, nor is there anything that can affect the time-geographic workings of society without affecting the path of an individual. This dialectic can be construed as subsuming various aspects of the preceding three dialectics. However, it also subsumes at least one additional relationship. Any time a person freely or involuntarily commits a segment of her daily path and finite time resources to an activity bundle of an institutional or organizational project, she subtracts from the total daily time resources of society as a whole in a given area. She thereby reduces the number of other institutional or organizational project-activity bundles that can be packed into the time–space organization of that same already defined portion of society. And, oppositely, every time the workings of society, or an institutional or organizational project and its activity bundles, increase (or decrease) the claim made on the daily (or biographical) time resources of an individual, it results in a reduction (or expansion) of the number of other activity bundles and interactions – and resultant internal experiences – that can be packed into her daily (or life) path.[21] Because of the individual–societal dialectic, and because the paths of all individuals are constantly in touch or in 'togetherness' with other people and objects, no person can either be born or die in a given place, or in-migrate or out-migrate from a given place, without adding to or subtracting from the societal-level accumulation of (actual and potential) activity bundles and inter-actions within that place.

The individual–societal dialectic, as well as the other time-geographic dialectics, take on added meaning if what is generally labeled by various observers and scholars as 'societal change' is regarded as always outwardly expressing itself as the introduction, disappearance or modification of projects usually defined by societal organizations or institutions. For, insofar as projects involve activity bundles and the linking up of individual paths, when 'societal change' is viewed in this manner it ceases to be an elusive abstraction and becomes inseparable from changes in the life content, or external actions and internal experiences, of individuals.[22]

Two overviews of individual behavior and its societal context

Using time geography as a foundation, at least two schematic
overviews of individual behavior and its societal context are
possible for the 'behavioral' and 'modern human' geographer.

One possibility is to modify an approach recently suggested by
Buttimer,[23] and to wed time geography to the *genre de vie* concept
of Paul Vidal de la Blache and his followers. As Buttimer points
out, Vidal's focus on *genre(s) de vie*, or the style(s) of living of
particular places or areas, represented an attempt 'to grasp the
ongoing dialectic of *milieu* and *civilisation*, the perennial tension
between the *milieu externe* (physically observable patterns and
processes) and the *milieu interne* (values, habits, beliefs and ideas)
of a civilization.'[24] If one can look beyond some of Vidal's ideological idiosyncrasies, the appropriateness of such a marriage is
suggested by Vidal's repeated contention that each *genre de vie* is
rooted historically in the 'habits and traditional social forms born
within the context of specific livelihoods,' his warnings 'about the
dangers of splitting up for analytical convenience those parts of
human and earthly reality that need to be understood in terms of
their coexistence in space and time,' and the classical concern of 'la
géographie humaine' to develop 'a keener sensitivity to one's own
life and milieux' (or one's own life and its societal context).[25]

This first schematic overview, quite simply, utilizes the language
of time geography to depict the existing (and emerging) *genres de vie*
of any given place or bounded area as parts of a structural whole.
This is done in terms that primarily emphasize individual–societal
relationships, but which also give some consideration to elements
and characteristics of the natural environment. At one focal point of
this schema (Figure 10.2) is the full mix of physically observable
genres de vie, or the local day-to-day, pulsating time geography; i.e.
the total choreography of daily and life paths danced out by
individuals (and the man-made objects and elements of the natural
environment that they employ) in fulfilling personally defined and
institutionally or organizationally defined projects of both a routine
and nonroutine nature. In keeping with the underlying dialectics of
time geography, the activity bundles and interactions that paths
move in and out of on the visible landscape are both shaped by, and
are the shapers of, the accumulated mental experiences of individuals (or their language, ideas, knowledge, memories, values,

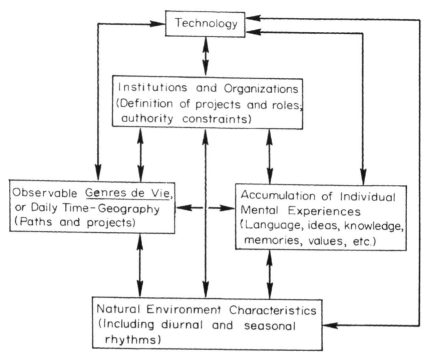

Figure 10.2   An overview of individual behavior and its societal context based on time-geography and *genre de vie* concepts. See text for a brief sketch of some, but not all, of the relations involved. Also compare and contrast with Figure 4.1 in Buttimer, 'Charisma and Context,' p. 68 (See note 20 for complete reference.)

attitudes, etc.). In this framework, the societal institutions and organizations (and their elites) that set authority constraints and define local projects (and independently existing roles) may be based either inside or beyond the place or area in question.

Technology also occupies a vital position in this schema: for, while being both the outcome of individual and cooperative research and experimentation based on accumulated experiences, and the expression of economic and other motivations built up at the individual and institutional (or organizational) levels through past paths and projects, it also affects, among other things, the ways in which many projects are defined (including what man-made objects and elements of the natural environment are to be included in certain activity bundles); the mobility, or capability constraints, of individuals; the tempo at which actions and interactions occur; and the content of mental experiences.

This unified structure of individual–societal relationships is not independent of the natural environment. The use of elements of the natural environment is synonymous with its transformation. Individual paths are constantly in touch and 'togetherness' with the natural environment. Most obviously, the characteristics of the natural environment (*including its diurnal and seasonal rhythms*, as well as its natural resources, topography, soils, vegetation, and climate), contribute both to the manner in which projects are defined and the mental experiences acquired by individuals.

It is not feasible in a literal sense to implement the scheme shown in Figure 10.2. A duplication of all the paths and projects of any bounded region is entirely impossible. However, with the unified structure of the time-geography-based schema in mind, fundamental insights into individual behavior and its societal context can be gained by deductively and humanistically working outward from both the attributes of a few representative projects and the incorporation (or elimination) of separate parts of those projects into the unbroken time–space paths of individuals.

A second, closely related overview recently proposed by Törnqvist also suggests the use of much the same methodology. Törnqvist's overview (Figure 10.3) focuses on the relationships between the 'registered behavior,' or project participation and path tracing, of particular individuals in the present (t_0), past (t_{-1}) and future (t_1). (The time scale adopted depends on the type of behavior observed.) His schema rests on four 'observation fields' or filters through which fragments of 'reality' are captured, plus the registered behavior itself, which is represented by a fifth field whose thickness reflects the fact that any form of project participation has a duration to it.[26]

The 'scope of action possibilities' within which project participation unfolds is constrained or otherwise influenced by: field A, the biophysical, or natural, environment; field B, the sociotechnical organization of the environment (a composite field subsuming 'all the kinds of organization invented by man in the financial, administrative and technical areas,' including physical infrastructure, formal organizations and their rules, codes and agreements, 'market forces,' etc.); and field C, the local transport systems that have been separated out from field B.[27]

Intermediating between 'registered behavior' and the 'scope of action' possibilities is field D, the 'environmental images' held by

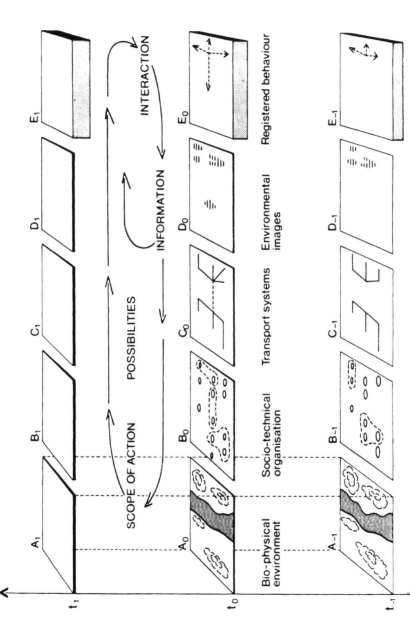

Figure 10.3 Törnqvist's 'observation field' overview of the 'registered behavior' of individuals and its societal context. Adopted from Törnqvist, 'Om fragment och sammanhang,' p. 261; and Törnqvist, 'On fragmentation and coherence in regional research.' (See footnote 18 for complete reference.)

individuals. These images refer to the individual's view of her situation, or her limited knowledge of and attitudes toward path alternatives. Such images, which are the result of the individual's previous path and project-participation history, have served as a basis for decisions that led to the presently 'registered behavior,' which, in turn, will influence future images and behavior. The cross-sections used in Figure 10.3 are merely a convenience, for Törnqvist emphasizes that he is dealing with a continual and complex process of change – a process that is lost sight of by specialized geographers and social scientists who attempt to 'explain' registered behavior E_0 on the basis of variables derived from A_0, B_0, C_0 or D_0, rather than seeing E_0 as being associated in not always directly apparent ways with conditions, events and experiences occurring at various points of time either between t_0 and t'_{-1}, or before t_{-1}.

It should be noted that both time-geography based overviews of individual behavior and its societal context are capable of bringing the social relations flowing from the means of production (and consumption) down to the individual in a very explicit way. Therefore they lend themselves, for those who so choose, to being grafted upon a marxist perspective. Because its unified structural view is somewhat more explicitly depicted, this is especially true of the schema summarized in Figure 10.2.

Before your path diverges

Before your path diverges, before you complete the project of reading this chapter, I would like to clarify a few points about time geography in general, and its application to issues surrounding individual behavior and its societal context in particular.

Time geography provides a means by which to integrate knowledge from different fields; a means by which to overcome the excessive, but not always unfruitful, fragmentation of knowledge in geography and the social and natural sciences relating to it; a means by which 'to restore a sense of wholeness and coherence and room for creativity within the life of the individual human person.'[28] By emphasizing 'that the local finitude of time and space makes everything hang together in a complex way,'[29] by allowing individual behavior and its societal context to be treated in the humanistic vocabulary of creative writing as well as in the vocabulary of simple deductive reasoning, time geography provides a *perspective* by which to grasp both the certain ambiguity of the internal and the ambiguous certainty of the external.[30] Perhaps most importantly, time geography is not to be seen as a new systematic specialty for geographers, but a general developing *perspective* 'anchored in certain basic facts of life,' which, because they are present in every human situation, 'one can possibly ignore or neglect, but hardly deny.'[31] Time geography is a way of viewing the world at large as well as the content of one's own life.[32] It is a 'developing perspective which at its core is disciplined like a formal model but at its front lies open like a living language.'[33] It is a perspective, like other not necessarily incompatible perspectives discussed in this book, which recognizes through its underlying dialectics that there can be no model of man that is separate from a model of society.[34] As yet, time geography remains 'the outline of a continent to be explored.'[35]

Notes

1 The position taken in these opening comments is not inconsistent with a basic tenet of the Frankfurt School of Critical Theory; to wit, that the particular can only be understood through reference to the whole and that the whole can be best illuminated through the study of its particulars. See, M. Horkheimer (1972) *Critical Theory: Selected Essays*, New York, Seabury/Continuum; P. Connerton (ed.) (1976) *Critical Sociology*, Harmondsworth, Penguin. On the place in geography of dialectical thinking – which is *not* confined to or synonymous with marxism – see B. Marchand (1978) 'A dialectical approach in geography,' *Geographical Analysis*, 10, 105–19.
2 G. Olsson (1978) 'Of ambiguity or far cries from a memoralizing mamafesta,' in D. Ley and M. S. Samuels (eds) *Humanistic Geography: Prospects and Problems*, Chicago, Maaroufa, 117.
3 T. Hägerstrand (1975) 'Ecology under one perspective,' in E. Bylund, H. Linderholm and O. Rune (eds) *Ecological Problems of the Circumpolar Area*, Luleå, Norrbottens Museum, 271.
4 See T. Hägerstrand (1974) 'On sociotechnical ecology and the study of innovations,' *Ethnologica Europaea*, 7, 33–4; T. Hägerstrand (1978) 'A note on the quality of life-times,' in T. Carlstein, D. Parkes and N. Thrift (eds) *Timing Space and Spacing Time, Volume 2: Human Activity and Time Geography*, London, Arnold, 214.
5 For example, see A. Pred (1973) 'Urbanization, domestic planning problems and Swedish geographic research,' *Progress in Geography*, 5, 1–76; A. Pred (1977) 'The choreography of existence: comments on Hägerstrand's time geography and its usefulness,' *Economic Geog-*

graphy, 53, 207–21; N. Thrift (1977) 'Time and theory in human geography: part II,' *Progress in Human Geography*, 1, 413–57; and the various items appearing in Carlstein, Parkes and Thrift, *Timing Space and Spacing Time*, 115–263 (see note 4 above).

6 Pred, 'Choreography of existence,' 208 (see note 5 above). For a diagrammatic clarification of capability, coupling and authority constraints, see T. Hägerstrand (1970) 'What about people in regional science?' *Regional Science Association*, 24, 7–21; and Pred, 'Urbanization, domestic planning problems and Swedish geographic research,' 37–47 (see note 5 above).

7 Hägerstrand, 'A note on the quality of life-times,' 220 (see note 4 above).

8 An exceptional circumstance arises when a telephone call is part of a project; for, while each of the parties must bring their paths in touch with a telephone, they need only coordinate their personal paths in time and *not in space* in order to form the requisite activity bundle. When there are a great number of activity bundles within a project they can usually be grouped into 'phases' whose number and kind should depend on the general character of the project in question.

9 Additional comments on the attributes of projects are contained in Hägerstrand, 'On sociotechnical ecology,' 17–34 (see note 4 above); K. Cederlund (1977) 'Administrativ verksamhet som projekt,' *Svensk Geografisk Årsbok*, 53, 69–80; T. Ek (1977) 'Energikostnader-transporter-arbetsorganisation: explorativa studier av förandringar och anpassningsmöjligheter inom svensk industri,' *Svensk Geografisk Årsbok*, 53, 81–92; L. Olander (1977) 'Företagsadministration som aktivitetssystem i tid och rum,' *Svensk Geografisk Årsbok*, 53, 55–68; L. Olander and T. Carlstein, 'The study of activities in the quaternary sector,' in Carlstein, Parkes and Thrift, *Timing Space and Spacing Time*, 209–12 (see note 4 above); A. Pred (1978) 'The impact of technological and institutional innovations on life content: some time-geographic observations,' *Geographical Analysis*, 10, 345–72.

10 T. Hägerstrand (1977) 'On the survival of the cultural heritage,' *Ethnologica Scandinavica*, 11.

11 It is important to realize that in highly industrialized societies the local chiseling out of niches for the activity bundles of new organizational or interorganizational projects involving production or distribution is often dictated either by the need to coordinate with other geographically distant activity bundles, or by the profit motives of far-away administrative units that control already existing local activity stations.

12 The specialized, independently existing roles occurring within organizations and institutions can be distinguished from one another in terms of the time-demanding routine and nonroutine project-activity bundles in which they obligate participation.

13 B. Lenntorp (1978) 'The time-compact society and its representation,' Uppsala, Paper presented at the Ninth World Congress in Sociology, 11.

14 In a related manner Cullen has recently formulated the following twofold argument. The 'everyday routine (in time–space) becomes a

manifestation of the long-term choice,' and it is *negative* responses to normal elements of the individual's daily activity pattern that usually 'trigger a new long-term choice.' I. G. Cullen, 'The treatment of time in the exploration of spatial behavior,' in Carlstein, Parkes and Thrift, *Timing Space and Spacing Time*, 27–38 (see note 4 above).

15 Note also, that to the extent successful post-childhood admission to independently existing roles depends upon an individual's inherited biological characteristics, such as strength or intelligence, it also not only depends on who her parents were, but exactly which egg and sperm were involved, or exactly when other projects allowed her parent's daily paths to join in the act of conception.

16 M. Godkin (1978) 'Rediscovering and rethreading identity in forgotten places: a phenomenology of the lived experience of alcoholism,' Unpublished paper, 2.

17 For remarks on time geography that can be related to this dialectic see A. Buttimer (1976) 'Grasping the dynamism of lifeworld,' *Annals of the Association of American Geographers*, 66, 277–92; C. van Paassen (1976) 'Human geography in terms of existential anthropology,' *Tijdschrift voor Economische en Sociale Geografie*, 67, 324–41.

18 See, C. Persson (1977) 'Omgivningsbilder och deras roll i beslutsprocessen: några reflektioner kring teori och metodik,' *Svensk Geografisk Årsbok*, 53; G. Törnqvist (1978) 'Om fragment och sammanhang i regional forskning,' in Expertgruppen for regional utredningsverksamhet (ERU), *Att forma regional framtid*, Stockholm, Liber Forlag, 246–83, also as *On fragmentation and coherence in regional research*, Lund Series in Geography, B, 45.

19 T. Hägerstrand (1976) 'Here and now as the waist of an hourglass,' Unpublished manuscript, 3 and 19.

20 Destruction due to storms or other natural causes can also be depicted in the path convergence–path divergence language of time geography, despite the absence of human paths and projects. See Hägerstrand, 'Ecology under one perspective,' 271–6 (see note 3 above).

21 See previous remarks about constraints on the time resources of individuals.

22 Recall also, from the discussion of the external–internal dialectic, that institutional and organizational projects depend upon the goals and decisions of individuals and thereby upon the course of their previous daily and life paths.

23 A. Buttimer, 'Charism and context: the challenge of La Géographie Humaine,' in Ley and Samuels, *Humanistic Geography*, 58–76 (see note 2 above).

24 ibid., 61.

25 ibid., 64, 73 and 69.

26 Törnqvist points out that a much greater number of 'observation fields' could be employed. 'Om fragment och sammanhang,' 252 (see note 18 above).

27 ibid., 252–4 and 280–2.

28 Hägerstrand, 'Here and now as the waist of an hourglass,' 30 (see note 19 above).

29 Hägerstrand. 'A note on the quality of life-times.' 224 (see note 4 above).
30 See. Olsson, 'Of ambiguity or far cries from a memoralizing mama-festa,' 109–20.
31 Carlstein, Parkes and Thrift, *Timing Space and Spacing Time*. Introduction, 121 (see note 4 above).
32 See, A. Pred (1979) 'The academic past through a time-geographic looking glass.' *Annals of the Association of American Geographers*, 69, 175–80.
33 Hägerstrand, 'Here and now as the waist of an hourglass,' 12 (see note 19 above).
34 See essays by D. Ley and K. R. Cox on pp. 209 and 256 in this volume.
35 Hägerstrand, 'Here and now as the waist of an hourglass,' 2 (see note 19 above).

11 Bourgeois thought and the behavioral geography debate*

Kevin R. Cox

The primary assertion of this paper is that behavioral geography, like location theory, cultural geography and even 'socially relevant' geography, is just one more instance of bourgeois thought: of the world conceived as a totality of things interacting in external cause-and-effect relations. This ontological and epistemological position is characterized as 'bourgeois' because, on the one hand, it is grounded in capitalist social relations experienced as things, and because, on the other hand, the practical implication of this 'bourgeois' consciousness is the preservation of capitalist social relations and, consequently, the characteristic appearances in which they present themselves to the world: bourgeois thought and capitalist practice mutually presuppose and produce one another.

The paper is divided into three major sections. In the first section, behavioral geography is characterized; this is carried out both in terms of its historical background within geography and in terms of the contribution behavioral geography has made to the discipline. In the second section, commonly encountered critiques of behavioral geography are discussed. These involve the neglect of the so-called social dimension and the assumption of subject–object separation. This leads into a discussion of humanistic geography as an articulation of both these critiques and as an escape from subject–object separation that is doomed to failure as a result of the

*I would like to acknowledge helpful discussions with Frank Z. Nartowicz of the University of Kansas prior to writing this paper.

ambiguities created by its inherent idealism. Finally we turn to dialectical materialism and *its* critique of subject–object separation as not some*thing* existing purely in the head, as the humanists would argue, but as grounded in the concretely experienced: the experience of social relations under capitalism as things and as relations between things.

Behavioral geography: a characterization

Historical background

The emergence of behavioral geography is often described as the behavioral revolution. A close examination of its immediate intellectual pedigree calls that characterization severely into question. Ontologically and epistemologically, behavioral geography represented in no way a radical break with the past. Rather, despite its pretensions, and as I will shortly demonstrate, behavioral geography was simply 'business as usual.' The changes it called for were cast in the context of current weaknesses of locational analysis and were designed to eliminate them, not through questioning philosophical presuppositions but through a simple incremental elaboration of existing behavioral postulates.[1]

Consider briefly locational analysis. It consisted, and continues to consist, of two major interlocking streams of thought. The first, in a temporal sense, is spatial analysis. This I will characterize as the pursuit of morphological laws. It has its roots in the search for the geographer's version of the Holy Grail: an object of analysis – in this instance space – that geographers can call their own. It allowed escape from environmental determinism only into another determinism of the spatial variety.[2]

The second stream of thought is location theory. This has its origin in neoclassical economics. For geography its relevance lies in disillusion with morphological laws as the goal of inquiry and in an enhanced awareness of the necessity for some theory to explain them and, indeed, to assist the search for laws. Concomitant with this was the sense that the Holy Grail had proved to be once more illusory and that space without process is no space at all. Locational analysis, therefore, rediscovered the human subject as the necessary agent of a pattern-creating process. Behavioral geography continues the elaboration of our knowledge of the subject with its space preferences, space learning, space perceptions, cognitions and the like.

In brief, with the necessary rediscovery of the subject exemplified by location theory, behavioral geography could not be far behind.[3] However, the emergence of behavioral geography is emphatically evolutionary rather than revolutionary. It is wholly a creature of a locational analysis that had become subject sensitive. Thus, behavioral geography has remained firmly embedded in the ontological and epistemological presuppositions of neoclassical economics, location theory, and, hence, in its revised versions, locational analysis.

Ontologically this is a view of the world as a whole or totality of thing-like subjects and objects interacting one with another in external cause-and-effect relations. In the neoclassical vision the subjective element is represented by the preference function, which has no ultimate determinant other than the individual himself; hence, in location theory and locational analysis, locational behavior can be seen as an outcome of *space preferences*. The objective element in neoclassical economics is represented by price ratios; hence the locational version of market behavior as locational adjustment to *price gradients*.

Subject and object interact in cause-and-effect relationships; what is produced (i.e. the objective) is a result of preference functions, albeit aggregated in the form of demand curves; and what is consumed (i.e. the realization of the subject) is determined by the objects produced and their attendant prices. Thus, as Gough has shown elsewhere,[4] in neoclassical economics and location theory 'the two philosophical strands of bourgeois thought, idealism and simple materialism, are triumphantly brought into harmony' by equating marginal subjective rates of substitution (i.e. the ratios of marginal utilities) to marginal objective rates of transformation (i.e. price ratios).

Epistemologically the rigid separation of subject and object is consistently maintained in the form of the assumption of the observer as separate from the observed. This has important implications for the conception of science and of scientific method.

The assumption of separation of observer from observed, for example, allows and makes plausible the view that facts and associations of facts are given to naive experience, that the social scientist is simply an observer and does not bring to bear upon the object of study any theoretical notions not ultimately derivable from that object of study in its naively given overtness.

Further, the assumption of subject–object separation contains within it the view that facts are separate from values, that the world of fact is unaffected by any personal values that the investigator might hold. Consequently the only reason investigators might disagree about the world of facts, holding observational methods constant, would be because of error originating in subjective bias. The problem, however, can be taken care of by the establishment of a mode of apprehension agreed upon by all investigators and permitting replication of observation and test.

In brief, the ontological assumption of subjects and objects as things interacting external in cause-and-effect relations receives a reflection at the epistemological level wholly consistent with the positivist project of generalization and prediction, and with its methodology, scientific method. Whether positivism presupposes these philosophical views as this argument might suggest, is, however, more contentious.

The contribution of behavioral geography

Consider now in a more specific manner the contributions to, and elaborations of, locational analysis, provided by behavioral geography. In this respect I think it fair to say that the initial *image* of behavioral geography was that it represented a relaxation, at the ontological level, of the assumption of subject–object separation. There *seemed* to be an attempt on the part of behavioral geographers to examine the object from the subject's point of view. Kirk's differentiation between a phenomenal and a behavioral environment, for example, nicely expressed the idea of the object as a cognitive, value-laden category.[5] It was this promise that explains a view which still persists with some commentators: the idea that behavioral geography represented an assertion of the role of subjectivity.[6]

This promise, however, has been belied by the actual course of events. The object, as it turns out, is not to be an object for a subject; rather, subject and object are to be kept apart, and in some versions of the behavioral credo the subject as subject is to be eliminated altogether.

Thus, with respect to the separation of subject and object the perceived or cognized environment becomes a perception or cognition as an attribute of, and intrinsic to, the subject. As such it can be, and is, compared with the object (i.e. the environment) 'as it really is,' confirming the subject–object dichotomy. Analytically, the percept, cognition, etc., as an attribute of the subject can be used in two ways:

1 As some*thing* with which to predict behavior. This appears to be the objective of work oriented to the identification of space preferences. The central idea is the notion of revealed preference borrowed from economic analysis; actual consumption behavior reveals a structure of preferences for goods that can be used predictively irrespective of the particular set of goods available. In space-preference studies, actual spatial behavior with regard to different attributes of locations (e.g. size of city, distance from consumer) is used as a basis for deriving a preference function that can be used to predict spatial behavior irrespective of the particular spatial context.[7]

2 As some*thing* to be explained by other objective attributes of the individual, such as socioeconomic status, or by the spatial environment. This type of work is pervasive in behavioral geography. It is typified by the following quote from a paper by Peter Gould on the acquisition of spatial information:

> *Where* people are located appears to determine, to a very large degree, what sort of information surfaces they will generate. Moreover, their levels of group agreement about certain, essentially geographic questions are also highly predictable from their locations. These locations, relative to the broad pattern of information-generating population in the country also shape the aggregate travel fields that represent possibilities of acquiring first-hand, as opposed to received, information.[8]

The same type of orientation is apparent in the following quote from Briggs; but, more importantly, quite why one would want to adopt this orientation is also disclosed:

> Finally we might ask: for what purpose is cognitive distance being examined? If our purpose is to understand the phenomenon *per se*, then it is critical to isolate independently the factors affecting it. Examining distances toward versus away

from the downtown is insufficient since the critical influents may be travel time, building intensity and variability, bias in the number of trips downtown, or any combination of these . . . . On the other hand, if our purpose is to find more realistic measures of spatial separation to use in predictive models of, for example, consumer movement, then our task is simpler. There is no need to determine the separate influence of factors which always covary spatially. As a geographer my purpose is the latter, but my measures of cognitive distance would rest upon a firmer foundation if the former could be achieved.[9]

Hence, much behavioral geography research, far from appearing to recognize and elaborate on the role of the individual as per its earlier image, seems to eliminate the individual as subject altogether; instead of the object being an object for a subject, the subject has become an object for an object that, by some weird alchemy, gets upon its legs and proceeds to act as a subject. This has been noted before. Cullen, for example, has written of behavioral geography: 'The primary orientation has never been the motivated behavior of the individual, but rather that individual's "mechanical" responses to spatial and social structures.'[10] Spatial determinism lives, therefore, and preserves continuities with past geographical work. To imply by that, however, that one can explain behavioral geography in terms of disciplinary traditions would be idealistic, and that is furthest from my intent.

Critiques of behavioral geography

In discussion of behavioral geography two critical comments are repeatedly offered: the neglect of the social; and the separation of subject and object. It seems to me that the neglect of the social is the less significant of these and can be discussed fairly briefly. The separation of subject and object, however, is by far the most fundamental and leads into a discussion of humanistic geography as an alternative to behavioral geography. Thus there are no inherently epistemological obstacles to an incorporation of the social into present work on behavioral geography. There is no way, however, in which a marriage could be arranged between behavioral and humanistic versions of geography.

Neglect of the social

Comment regarding the necessity to pay greater attention to so-called social factors or the sociological dimension is common:

> An example of the literature on spatial behavior suggests that one critical missing link is the sociological dimension. Most of the explanatory models rest heavily on generalizations, about relationships of organisms to their environments; for example, perceptual/cognitive processes; . . . the sociological dimension in these processes is rarely given explicit attention. Similarly, life style, social stratification, status and role are rarely treated explicitly in studies of environmental behavior.[11]

Quotes of this character could be multiplied many times.[12] Of course it is easy to exaggerate the degree to which a sociological dimension in this sense has been missing; the concept of a social communication network, for example, has been central to Hägerstrand's work on diffusion and migration. Moreover, this is a body of work that has stimulated research elsewhere in behavioral geography.[13] Nevertheless, by and large the point is well taken.

Not that we should be surprised by this omission. After all, in the views of its participants behavioral work was based on the conviction that 'explanations of spatial events must ultimately be related to the individuals initiating those events,'[14] i.e. the elaboration of the subject referred to above. However, there is no fundamental reason of an epistemological character why Buttimer's missing link could not be inserted within the existing framework of behavioral geography. Hägerstrand has indicated one direction in which this might go, and there is a lively complementary literature on the development of social networks. Likewise the neoclassical assumptions from which location theory and ultimately the greater part of behavioral geography derive have shown themselves capable of yielding interesting statements about the development of social life and power relations. The public-choice literature is to the forefront here[15] and has found some reflection in the work of Wolpert;[16] but there is other ingenious work on, for example, social interaction as a micro-economic problem.[17]

The separation of subject and object

The most important criticism has been reserved for the ontological and epistemological assumption of subject–object separation. The central ontological point is the subjectivity of the objective; the object is always an object for a subject. Man, it is argued, has the unique capacity of self-awareness: he can separate himself from his acts. As such he is able to reflect on those acts, construe, plan, impute meanings, act and learn from his actions. Intervening between the objective world and behavior is consciousness, which imposes its own interpretations upon the objective world and thus independently affects behavior. Man, therefore, cannot be understood in terms of the frozen categories of 'being;' the individual, according to this view, is not another 'fact' given to naive experience. Rather he is in a constant state of 'becoming,' certain of nothing but uncertainty. Programmatically, Olsson has stated that

> our real need is for a new philosophy of man. In that philosophy the human condition will be conceived as an evolving relationship between the contradictory forces of certainty and ambiguity, man and men, men and society, freedom and necessity.[18]

The epistemological implications of this view are quite far reaching. Firstly, if one grants the basic ontological point of the subjectivity of the objective, the factual is no longer given to naive experience. The factual becomes merely the meaning the investigator attributes to a particular event, and may be substantially at variance with the meanings imputed by the investigated. This has two important effects. The first is to undermine the possibility of scientific laws of a universal-truth status. Events in different cultures, for example, may *appear* the same to the investigator, but have quite different meanings and consequently different consequences. Likewise, as meanings shift over time, in the process of 'becoming,' so will the 'facts' have different meanings to the actors involved – but not necessarily to the investigator reliant on external appearance and his own theories, who imposes his own meanings on those appearances. Moreover, the transformation of meanings over time will be in part a result of the findings of scientific investigation itself providing another manifestation of that self-consciousness that undermines the possibility of social-science laws of universal-truth status; for human beings 'may become "involved" in the process of social science in the sense that they are capable of standing back from a commentary upon their performance and initiating an action which is deliberately designed to confound any implicit prediction.'[19]

The second effect is the related one of problematizing the investigator's frame of reference. In behavioral geography the focus is on behavioral forms as objects given to naive observation: the frame of reference of the investigator is, therefore, nonproblematic. This structures methodology. The necessary assumption given the necessarily nonproblematic frame of reference of the investigator is that the particular categories used for gathering data have the same meaning to the investigator as to the investigated. However, given that objects are not given to naive experience but are objects for subjects, the investigator's frame of reference is called severely into question.

This is seen most dramatically in survey research. For example, the categories supplied by the investigator may have quite different meanings to different individuals in the sample. Likewise the responses supplied by the investigator on the assumption that they are exhaustive may be far from so. As a consequence one may learn more about the behavior of the sample in responding to a set of categories the investigator attempts to impose upon them, than about the behavior under investigation itself.[20]

There is in addition a second major epistemological implication: the inseparability of facts and values. If objects are objects for a subject and, therefore, are construed with respect to the intentions and needs of the subject, then facts are also necessarily stained with the evaluatory. Definitions of the world must be consistent in some way with the needs and interests of the subject.

In a practical, everyday sense this is obviously valid; the Burmese peasant, for example, can differentiate far more types of bamboo than can the European, simply because he is dependent on bamboo for a variety of needs extending beyond those of his tomato plants. On the other hand, the separation of manual and intellectual labor makes this a less plausible comment on self-consciously scientific work. Yet, if we concern ourselves less with the facts particular to a culture and more with the *conception* of 'fact' held in social science, it becomes clear that there is a problem.

In the positivist world facts are given to the senses – 'speak for themselves' – in a manner that is nonproblematic. As a result it is possible to arrive at generalizations and predictions about the world. The form of my argument here suggests that it is the conception of 'fact' that produces the possibility of generalization and prediction. Might it not be, however, that it is the desire for a particular form of knowledge, i.e. knowledge that allows prediction – and consequently control – that governs the positivist conception of 'fact'? Instead, therefore, of generalization and prediction being mere afterthoughts subsequent to the discovery of a world of universally true things, the conception of the world as one of facts interacting in cause-and-effect relations may be a function of a particular set of social values regarding technical and social control. It may well be, therefore, that it is values that promote the 'thingified' view of the world central to the positivist view.

Humanistic geography

One significant, and indeed important, intellectual response to these problems is the call for a humanistic geography. This may be discussed in terms of two dominating ideas, one of them primary and the other of secondary significance. Primary and central to the view is the assumption of anthropocentricity. This asserts that the world is first and foremost the world of self-conscious individuals. Objects presuppose subjects. It is a set of meanings imposed on the world in accord with individual values, intentions and subjective constructions that structure actions expressing those intentions. Hence, elucidation of place commences with a reconstruction of the meanings the subject has developed and imposed on place. And reconstruction of meanings necessitates that the investigator bracket his categories and get through to pure experience: 'This method does require imagination and an ability to push one's own theories to the back of one's mind.'[21]

A somewhat secondary role is assigned to the idea of the social nature of experience. The construction and imputation of meaning takes place within social worlds that facilitate convergence in values, meanings and definitions of the situation.[22] Subjectivity is muted, therefore, by the fact that the '(human) life world is an inter-subjective one of shared meanings of *fellow men* with whom he engages in face-to-face *we* relationships.'[23] Buttimer has made essentially the same point though in a manner leaving less space for subjectivity.[24]

This secondary element in the humanistic view provides a useful point of entry into a discussion of its abiding weakness: for, an expression – but only an expression – of this weakness is this highly impoverished view of the social. Social life is assumed rather than explained. Indeed, it is almost as if it has been inserted in order to allow for some convergence in individual meanings and recreate the possibility of generalization.

In brief, the humanistic viewpoint has no theory of social life adequate to the task of clarifying meaning. It provides no understanding of the forces that bring men together and which, therefore, also divide them. Hence it is curiously silent about the forces molding the values and intentions so important for the meanings the humanist puts at the center of the picture. It has no theory, other than that of communication, to explain the meanings men arrive at or to explain the tortuous paths they tread in the quest for meaning. Nor yet, therefore, can it have any notion of the forces motivating some to the construction of an environment molding the meanings of others – apart, that is, from a retreat into the idealism of a technological rationality[25] or an entrepreneurial ethic. The world of the humanist is without a dynamo for bringing men together, orienting them towards particular problems and particular solutions.

The reason is that it neglects the practical problems in answer to which social life must be reproduced and around which individuals become oriented according to their disparate practical interests. As Gregory has noted, explanation in terms of the subjective meanings of the actions of human beings is important but not sufficient;[26] there are certain constraints on action that are so taken for granted that the actor is unable to verbalize them or would regard it as irrelevant to do so.

This has important consequences, not the least of which is the theory of knowledge implicit in the humanist position. In its conception of the emergence of schools of thought, for example, there is an emphasis upon the subjective basis for adopting particular stances and the channeling of subjectivity in limited social worlds,[27] of which academic disciplines or departments might be prototypical. There is, therefore, no sense of relationship to real problems immanent in the development of society.

Thus the humanistic perspective is not able to answer within its own terms the naturalistic epistemology of social science to which it is so averse. Indeed, wrenched loose from its moorings in the practical interests stemming from the imperatives of social reproduction, the naturalistic epistemology of subject–object separation assumes the form of a pure idea that can be abolished by an act of will. Buttimer, therefore, remarks that the 'predominantly Cartesian view of knowledge' constraining social scientists, fosters a peculiarly managerial perspective on urban life. The solution, apparently, lies in 'frameworks for investigation and reflection which do not segment and ossify parts of the city as do Cartesian practices' and in 'an empathetic understanding of urban life as existential reality.'[28] Yet the practical feasibility, in political and ideological terms, of these proposals goes unremarked.

The fundamental problem, of which the inability to theorize social life is but one, though an important, expression, is the idealism inherent in the humanist position. While the object presupposes a subject, the subject does not presuppose an object. There is in humanistic geography a built-in tendency to ignore the material presuppositions of the ideas, intentions and subjective constructions that it seeks to emphasize, an inability to see that ideas are products of the brain sensorily engaging the material world and not deriving from the brain in some way independent of, and indifferent to, material existence.

Humanist geographers are not insensitive to criticisms of this nature. As David Ley himself has written: 'Without an environment we have only consciousness, albeit collective consciousness, expressing its will in a flaccid world relieved of the brute reality of material existence.'[29] However, there are only two ways of conceptualizing the relation of object to subject (as opposed to the relation of subject to object), and both of them lead away from humanistic geography in quite different directions.

On the one hand, it is possible to conceptualize the relation as an external one. The environment, for example, impinges on the individual's consciousness in an external, indifferent, thing-to-thing-like manner. This, however, gives way to an ontology of subject–object separation, the definition of consciousness and meaning as yet additional explanatory variables, and that vacuous pluralism that is but one more hallmark of bourgeois thought. Indications of this particular reaction to the accusation of idealism are numerous. Cullen, for instance, in answering his critics, seeks refuge in the idea of 'redressing a balance – a balance which has

tilted so far in the direction of mechanical and structural explanations that it is in danger of falling off the pivot,' and in 'sensible,
middle-of-the-road position(s).'[30] Ley, likewise, in his paper in this
volume, talks of 'a growing consensus in the social sciences . . . that
adequate explanation must include both the fatalism of social
structures and the creative spontaneity of the life world.'[31]

On the other hand, instead of viewing the relation of the object to
the subject as external, one can view it as an internal relation; the
object not only presupposes a subject, therefore, but the subject
presupposes an object: forests presuppose foresters, as the
humanist would assert, but equally foresters presuppose forests.
But this would mean that humanistic geography had turned its back
on bourgeois thought. It is in this direction that we must now
proceed.

Dialectical materialism[32]

The unity of subject and object

Central to dialectical materialism is the unity of subject and object.
Subject as purpose and object as condition for realizing that
purpose mutually presuppose and produce one another: just as
tilling as purpose presupposes tillable fields as an objective condition, so tillable fields presuppose tilling as purpose. The subject–
object relation, therefore, is, simultaneously, a production
relation, a property relation and an internal relation. A particular
purpose *contains* particular objective conditions and vice versa:
particular purposes are only possible and intelligible given particular objective conditions, and particular objective conditions are
only meaningful and feasible given certain human purposes. What is
important in viewing society is neither the subject nor the object,
but the internal relation linking the two; or, which is the same thing,
subject and object seen as internal relations.

Thus, the concepts through which the individual becomes aware
of himself and his world are grounded neither in the subject nor in
the object but in the relation of subject to object. Consciousness is
grounded in, presupposed by, particular subject–object relations.
There is a unity of theory and practice. Categories have presuppositions in subject–object relations without which they would
not be valid. The category of greed, for example, is only possible

where subject–object relations are mediated by money and, hence, where consumption is not limited by (e.g.) the variety of products currently available or by the problems of storing them.

Just as concepts are grounded in subject–object relations, so are individuals. Individual acts are presupposed by subject–object relations that make acts possible and meaningful. Although in our thingified view of the world we tend to view such properties as 'greed' and inherent in individuals, it is on closer inspection seen to be an internal relation presupposed by subject–object relations as noted in the paragraph above. Likewise an individual can be industrious, lazy and unimaginative only in a mode of life constituted by subject–object relations that make those activities possible and comprehensible. To be a writer, for example, presupposes a separation of mental from manual labor, the development of language in its written form, printing, etc.

Just as there is a unity of theory and practice so there is a unity of individual and society. As Marx wrote:

> The human being is in the most literal sense a political animal, not merely a gregarious animal, but an animal which can individuate itself only in the midst of society. Production by an isolated individual outside society – a rare exception which may well occur when a civilized person in whom the social forces are already dynamically present is cast by accident into the wilderness – is as much of an absurdity as is the development of language without individuals living *together* and talking to each other.[33]

This view of subject and object as internal relations contrasts with the view of positivism of the separation of subject and object. In this case neither subject nor object are defined with respect to each other. Objects and subjects are seen as things given to the senses and accountable in terms of external relations of cause and effect with other things. The subject has no necessary control over the object and vice versa: interactions of subjects and objects are accidental rather than presupposed, indifferent rather than necessary; tilling as purpose does not presuppose tillable fields as objective condition and nor does the attribute of a field as tillable presuppose tilling as a purpose. In other words, subjects and objects and concepts are ungrounded; their history as social creations is denied.

Society as totality of external relations

Understanding of the positivistic view is provided by dialectical materialism, the grounding of consciousness in subject–object relations. Theory and practice form a unity such that they presuppose one another: practice is the object through which a theoretical subject realizes itself just as a theoretical subject presupposes a particular practice. Yet practice is, itself, an internal relation of purposeful subject to objective conditions. To understand the theories, the consciousness characteristic of a particular mode of life, therefore, we must understand the subject–object relations that presuppose them; but not as they actually are, rather as they *appear* to the subject involved and as they are experienced.

Marx argued that, under capitalism, the subject experiences the objective conditions of realizing its purpose as separated from itself. Objective conditions are not experienced as presupposed to the purposes of the human subject, and as its own. The relation of subject to objective conditions is experienced as accidental and beyond the subject's immediate control. Marx pursues this idea through the concept of alienation, meaning, literally, separation. We can illustrate this important idea as follows:

1 *Alienation from activity as an objective condition:* The subject, by selling its labor power to capital, to dispose of it as capital sees fit, relates to its activity as an objective condition for another; not, therefore, as an objective condition presupposed by the laborer's purpose. Activity is some*thing* the laborer relates to accidentally depending on whether or not he can be employed: control of activity as an objective condition for another is assigned to that other.

2 *Alienation from the means of production as objective conditions:* The subject experiences the means of production not as presuppositions for his subjectivity but as presuppositions for the subjectivity of another – the capitalist. The means of production, since they are under the control of the capitalist, are seen as some*thing* the laborer relates to accidentally.

3 *Alienation from fellow men as objective conditions:* The subject experiences others not as presuppositions for his subjectivity, not as the internal relations that make him the individual he is, but as external, indifferent, relating accidentally according to the dictates of the market, as an obstacle to the realization of

his purposes. Men know each other not as social beings whose needs demand mutual cooperation, but as private and competing entities. Thus the social relationship is experienced as external, as means to an end, and this experience is historically grounded: 'Only in the eighteenth century, in "civil society" do the various forms of social connectedness confront the individual as mere means towards his private purposes, as external necessity.'[34]

Instead of the unity of individual and community, individuals and communities are conceived as things apart. Language, social status, income, behavior are conceived as objective attributes of the individual rather than grounded in the subject–object relations that actively constitute a community. Congruent with this conception of natural, presocial individuals, community is experienced as a product rather than as a presupposition: a thing in the diverse forms of state, labor union, voluntary organization and family, with which the individual relates in a cause-and-effect manner – causing the state to do this and being caused by the labor union to come out on strike, for instance.

Hence, one arrives, through experience, at a conception of society as a totality of external relations: capital, labor, foreign labor, domestic labor, community, government, all interacting one with another in cause-and-effect relations. In this way capital can be conceived as job creating and causing the welfare of labor (note 'welfare' as another objective attribute), with communities competing one with another and affecting each other's chances of causing an inflow of job-creating capital and the like.

The commodity relation

Yet in a sense this only poses the problem afresh. If one apprehends the world as an interaction of self-sufficient things as a result of the separation of subject from object, whence the separation? For Marx, the critical consideration was the production relation defining the capitalist mode of production: the object as commodity. As a production relation, the commodity relation is necessarily a property relation – a subject–object relation: the object as the exclusive yet transferable possession of the owner.

As commodity the object – means of production, activity, product – assumes not only a use value with respect to a subject but also an exchange value: it enters exchange taking its exchange value with it, an exchange value, moreover, *seemingly independent* of its value in use. For the subject as seller, for example, the commodity has no use value, only a value in exchange.

In its completely developed form commodity exchange presupposes and is presupposed by the separation not only of producer from product but also of subject from the means of production. All the relationships of production are now mediated by a subject–object relation assuming a tangible form, i.e. the money relation that includes the capital relation. In such a form exchange values *appear* sprung loose, liberated from the tangible use values upon which they are, however, ultimately predicated.

For the subject, the objective conditions now appear as exchange values in the material form of capital and money. As exchange values seen as divorced from use values, they are experienced as things independent of activity and hence subjectless, relating to the subject in only a contingent, accidental fashion. It is to the exchange values that the subject necessarily orients, since use values are dependent upon entering into exchange. Yet as socially determined exchange values, commodities appear as separate, alien, thing-like and autonomous – beyond individual control. By relating to the objective conditions as exchange values – prices, wages, rents, profits – the subject surrenders *mastery of* the objective conditions and yields himself to *mastery by* the objective conditions. Relating to the objective conditions now assumes a contingent form – 'Is the price right?' – rather than the necessary form of mutually presupposing subject and object implied by the objective conditions as use values.

Thus labor becomes the condition for the realization of the objective conditions as values for themselves, realization in the form of a rate of return. Labor appears as a thing with which capital relates as a precondition for realizing itself as exchange value.

But underneath the appearance of separation lies a unity, a mutual presupposition of subject and object, of theory and practice and of individual and society, which capital, much as it would like to forget, can never escape, and of the existence of which it is periodically and painfully reminded. For, under capital and the commodity form, exchange value and use value presuppose one another; to have an exchange value, a commodity must have a use value; and to have a use value, it must also have an exchange value. For example, for exchange value to be realized by the capitalist the commodities produced must have a use value for consumers; and for the laborer's activity to have a use value for labor it must be realized in exchange.

But, paradoxically, exchange value is pursued as a thing apart, divorced from its necessary presupposition in use value. Hence, unity is denied and the contradictory character of the capitalist mode of production consequently exposed to view. It is exposed to view, for example, in overproduction. In the pursuit of exchange value through maximizing production, capital forgets that exchange values cannot be realized unless they are also use values.

A divided object presupposes and is presupposed by a divided subject. While the object is seen as separated into use value and exchange value, the subject separates into labor and capital. For labor, the ultimate purpose is to obtain use values, yet, in order to obtain use values, labor must first realize itself as exchange value. For capital, the ultimate purpose is to obtain exchange values: capital treats labor as a thing, an external relation, a mere means to the realization of exchange values. Equally, labor, until it learns better, treats capital as a thing – a means to the realization of use values.

However, by treating labor as a thing, capital reveals that its purposes are not identical with those of labor and that capital represents a – the – obstacle to the achievement of use values. This is evident, for example, in the unemployment that accompanies periodic slumps resulting from capital's failure to realize exchange values; or in unemployment created by the substitution of dead labor for living labor in the attempt to minimize production costs and so realize exchange value; or in the tendency for the price of labor to fall below the value of labor.

This is not to suggest that capital has any alternative to treating labor as a thing: the realization of exchange value necessitates it. But as a consequence so does the commodity relation necessitate a division of interest between capital and labor. In order to realize use values, labor must, and will, organize and forcibly remind capital of the indissoluble unity of use value and exchange value.

Labor's tendency, as labor rather than as not-labor, is to treat capital much as capital treats labor – as a thing. Thus Marx defines labor unions as 'insurance societies . . . formed by the workers themselves for the protection of the value of their labor power.'[35] By bargaining with capital as a thing, therefore, labor confirms its definition by prevailing economic science as a factor of production. This thingified view of the world is equally evident in the liberal reforms pursued by labor and adopted by the state.

But by so doing labor merely reproduces its relation to capital as a thing apart, the irreconcilable difference of purpose between labor and capital. Only as labor discovers this, as discover it it will, does it begin to realize that the obstacle to achieving use values is indeed this thingified conception of the world grounded in a separation of subject from object, of purpose from objective conditions, of labor from its objective conditions in the form of surplus value appropriated by the capitalist, and, therefore, capital itself.

Conclusion

From the standpoint I have developed in this paper behavioral geography now appears as pure ideology, and status quo ideology at that. It grounds its concepts and theories in the world of appearances. By so doing it accepts the normality, the total 'given-to-the-senses' naturalness of that world and the appearance of separation in practice underlying that conception. If locational analysis thingified social relations as spatial relations, behavioral geography ossifies them as spatial behavior, space perceptions, cognitions and the like. And, in this way, behavioral geography does its part, inadvertent as it might be, in preserving the status quo and in obstructing more radical and revolutionary conceptions of human practice.

What I have argued is that positivism is grounded in the way capitalism appears. Dialectical materialism, on the other hand, remembers the underlying unity of subject–object relations torn asunder by capital, and so must necessarily be a critique both of capitalism, of logical positivism and, consequently, of behavioral geography.

And just as behavioral geography has failed to escape from the straitjacket of subject–object separation, so, oddly enough, have the critics. The call for the incorporation of a sociological dimension into behavioral geography, for example, unwittingly accepts the assumption of the separation of individual and society that we have seen to be symptomatic of the problem. In the humanist view, for example, society is mere gregariousness affecting in a cause-and-effect manner the meanings imputed by the individual. Yet, as Marx argued, 'the human being is in the most literal sense a political animal, not merely a gregarious animal.'[36] Men interact and influence each other's views. But why they interact and should want to influence each other is, in the humanistic view, left suspended – suspended, that is, from the presuppositions of social life in necessary subject–object relations through which men reproduce themselves. Given this suspension, the naturalistic notion of gregariousness (i.e. the individual as separate from society) is the only feasible explanation. But, as Marx asserts, man is a political animal because, in order to realize his purposes, he must master the objective conditions of his existence.

Likewise the humanist assertion of subjectivity ultimately founders on the same shoals. Sensitive to the accusation of idealism, the humanists are forced to try to steer a course between the Scylla of subject–object separation and the Charybdis of subject–object unity. On the one hand they assert the subjectivity of the objective, on the other hand the objective lives. David Ley, for example, has to agree that:

> the environment is not in the head. Consciousness cannot break loose from a concrete time–space context, from the realities of everyday living; notions of pure consciousness are as much an abstraction from human experience as any isotropic plain.[37]

Yet the object cannot be both objective *and* subjective.

The tendency inherent within this contradiction has been to drive humanistic geography towards 'a synthesis that appropriately links man and environment, human intentionality and structural factors,'[38] and so into the stale possibilist–determinist debate that has forever bedevilled geography, and which will continue to do so, so long as the discipline is rooted in a conception of the world as a world of things – including consciousness as a thing. Only when we realize that consciousness is not a thing but a subject–object relation, and that there is a unity of theory and practice, will we escape these debates of which the debate over behavioral geography is but the most recent example.

Notes

1 D. Harvey (1970) 'Behavioral postulates and the construction of theory in human geography,' *Geographica Polonica*, 18, 27–45.

2 Tobler's famous 'first law of geography' is exemplary: 'everything is related to everything else, but near things are more related than distant things.' W. R. Tobler (1970) 'A computer movie simulating urban growth in the Detroit region,' *Economic Geography*, 46, 236. The most self-conscious statement of the morphological position is, of course, that of W. R. Bunge (1966) *Theoretical Geography*, Lund, Gleerup.

3 G. Olsson and S. Gale (1968) 'Spatial theory and human behavior,' *Regional Science Association*, 21, 229–42.

4 A. J. Gough (1976) 'Social physics and local authority planning,' in M. Edwards *et al.* (eds) *Housing and Class in Britain*, London, Political Economy of Housing Workshop, 87–104.

5 W. Kirk (1963) 'Problems of geography,' *Geography*, 48, 357–71.

6 For example, consider the following quote from R. Rieser (1973) 'The territorial illusion and behavioral sink: critical notes on behavioral geography,' *Antipode*, 5, 52:

> in line with the emphasis on the study of process came the need to examine the views of the environment held by different groups and individuals – their images (Boulding, 1956), mental maps (Gould, 1967), or more generally the cognitive milieu (Sprout, 1965) or behavioral environment (Kirk, 1951). These approaches were, broadly speaking, phenomenological: the attempt was to try and see the world as the individuals or groups concerned actually saw it, an approach well expounded by Lowenthal (1961).

The references he quotes are: K. Boulding (1956) *The Image*, Ann Arbor, University of Michigan Press; P. R. Gould (1966) 'On mental maps,' Michigan Inter-University Community of Mathematical Geographers, Discussion Paper 9; H. Sprout and M. Sprout (1965) *The Ecological Perspective on Human Affairs*, Princeton, Princeton University Press; W. Kirk (1951) 'Historical geography and the concept of the behavioral environment,' *Indian Geographical Journal*, Silver Jubilee Volume, 152–60; D. Lowenthal (1961) 'Geography, experience and imagination: towards a geographical epistemology,' *Annals of the Association of American Geographers*, 51, 241–60.

7 G. Rushton (1969) 'Analysis of spatial behavior by revealed space preference,' *Annals of the Association of American Geographers*, 59, 391–400.

8 P. R. Gould (1975) 'Acquiring spatial information,' *Economic Geography*, 51, 99.

9 R. Briggs (1973) 'Urban cognitive distance,' in R. M. Downs and D. Stea (eds) *Image and Environment*, Chicago, Aldine, 388.

10 I. G. Cullen (1976) 'Human geography, regional science and the study of individual behavior,' *Environment and Planning*, A, 8, 405.

11 A. Buttimer (1972) 'Social space and the planning of residential areas,' *Environment and Behavior*, 4, 285.

12 T. Hägerstrand (1970) 'What about people in regional science?' *Regional Science Association*, 24, 8; D. Mercer (1972) 'Behavioral geography and the sociology of social action,' *Area*, 4, 48; J. Wolpert and R. Ginsberg (1969) 'The transition to interdependence in locational decisions,' in K. R. Cox and R. G. Golledge (eds) *Behavioral Problems in Geography: a Symposium*, Evanston, Ill., Northwestern University, Department of Geography, Studies in Geography 17, 78.

13 An example is that of studies in voting geography where the geographical pattern of partisan preferences is interpreted, at least in part, as a function of flows of information and influence through a distance-biased communication network. See P. J. Taylor and R. J. Johnston (1979) *Geography of Elections*, Harmondsworth, Penguin.

14 D. Amedeo and R. G. Golledge (1975) *An Introduction to Scientific Reasoning in Geography*, New York, Wiley, 6.

15 G. Tullock (1970) *Private Wants, Public Means*, New York, Basic Books.

16 J. Wolpert *et al.* (1967) 'Coalition structures in three person, non-zero-sum games,' *Papers, Peace Research Society*, 7, 97–108.

17 G. Becker (1974) 'A theory of social interactions,' *Journal of Political Economy*, 82, 1063–93.

18 G. Olsson (1974) 'The dialectics of spatial analysis,' *Antipode*, 6, 59.

19 Cullen, 'Human geography, regional science and the study of individual behavior,' 407 (see note 10 above).

20 H. Braverman (1974) *Labor and Monopoly Capital*, New York and London, Monthly Review Press, 28, reports a classic case of this:

> in the many polls conducted according to the conceptions of W. Lloyd Warner – by Gallup, by *Fortune* in 1940, etc. – in which the population is classified into 'upper,' 'middle' and 'lower' classes, and into subgroups of these, vast majorities of up to 90 percent predictably volunteered themselves as the 'middle class.' But when Richard Centers varied the questionnaire only to the extent of including the choice 'working class,' this suddenly became the majority category by choice of the respondents.

As Braverman continues, 'Here we see sociologists measuring not popular consciousness but their own.'

21 L. Guelke (1974) 'An idealist alternative in human geography,' *Annals of the Association of American Geographers*, 64, 198.

22 Mercer, 'Behavioral geography and the sociology of social action' (see note 12 above).

23 D. Ley (1977) 'Social geography and the taken-for-granted world,' *Transactions of the Institute of British Geographers*, NS, 2, 505.

24 A. Buttimer (1972) 'Social space and the planning of residential areas,' *Environment and Behavior*, 4, 285–6.

> If environmental behavior is taken as the external (spatial) expression of social reference systems (sociological spaces) it becomes possible to integrate findings from the various levels of analysis . . . . The reference groups from which an individual derives his values and behavioral norms dictate [*sic*] certain aspirations and attitudes towards his milieu.

25 E. Relph (1976) *Place and Placelessness*, London, Pion.
26 D. Gregory (1978) *Ideology, Science and Human Geography*, London, Hutchinson, 125.
27 Ley, 'Social geography and the taken-for-granted world,' 504 (see note 23 above).
28 Buttimer, 'Social space and the planning of residential areas,' 281 (see note 24 above).
29 D. Ley (1978) 'Social geography and social action,' in D. Ley and M. S. Samuels (eds) *Humanistic Geography: Prospects and Problems*, Chicago, Maaroufa, 51.
30 I. G. Cullen (1976) 'The "new" behavioral geography – some comments upon the letter from Sayer and Duncan,' *Environment and Planning*, A, 9, 234.
31 Ley, 'Behavioral geography and the philosophies of meaning,' in this volume, p. 209.
32 The view of dialectical materialism presented in this section may be surprising to some. It is certainly not the dialectical materialism associated with more structuralist readings of Marx. Rather it is the blending of Marx's earlier humanism and his critique of political economy which comes to fruition in the *Grundrisse*, Harmondsworth, Penguin Books, 1973, but of which there are earlier signs in *The German Ideology*, New York, International Publishers, 1970. Regarding *The German Ideology*, see D. Sayer (1975) 'Method and dogma in historical materialism,' *Sociological Review*, 23, 779–810. On the reading of the *Grundrisse*, see R. W. Bologh (1979) *Dialectical Phenomenology: Marx's Method*, London, Routledge & Kegan Paul, which has been particularly helpful in clarifying my thoughts not only for this section of the paper but for the remainder as well.
33 Marx, *Grundrisse*, 84 (see note 32 above).
34 ibid., 84.
35 K. Marx (1976) *Capital, I*, Harmondsworth, Penguin, 1070–1.
36 Marx, *Grundrisse*, 84 (see note 32 above).
37 Ley, 'Social geography and social action,' 45 (see note 29 above).
38 ibid., 52. Cullen attempts to negotiate the problem by introducing spheres of relevance in the form of macro- and micro-scales of analysis for the object as objective and the object as subjective respectively:

> no single approach is appropriate at all levels of aggregation . . . . There is a sense in which social institutions persist and affect the lives of all of us. Many of them have important spatial implications. And many are so persistent as to warrant the use of mechanistic predictive models. What the macrotheorist must avoid, however, is

the tendency to assume that his models of macrophenomena have deterministic implications at the microlevel. And what the behavioral geographer must avoid, even more assiduously, is the tendency to assume the same thing. (Cullen, 'Human geography,' 408 (see note 10 above).)

Here consciousness is a thing operating at the microscale while the objective operates at the macrolevel. The slide into the determinism–possibilism debate is again very evident.

Author index

Abelson, R.P., 91 n.13
Adams, J.S., 175 n.56, 188, 192, 202 n.23, 203 n.41
Adler, T., 178 n.96
Agnew, J., 141 n.13
Ajzen, I., 143 n.46
Alonso, W., 172 n.5, 202 n.28
Ambrose, P., 175 n.55
Amedeo, D., 15 n.5, 141 n.13, 228 n.17, 227 n.14
Archer, J.C., xxviii n.24
Arsdol, van M., 189, 202 n.24
Atkinson, R.C. 57, 65 n.35
Austin, G.A., 42 n.19
Avery, K.L., 200 n.13, 291 n.14

Barnum, H., 63 n.3
Barrett, F.A., 204 n.63
Bassett, K., 193, 203 n.46
Baumol, W., 173 n.30
Bechdolt, B., 172 n.8
Beck, R.J., 165, 180 n.126
Becker, G., 277 n.17
Ben-Akiva, M., 178 n.96
Berger, P., 223
Berlyne, D.E., 42 n.22
Berry, B.J.L., 55, 63 n.4, 90 n.3, 141 n.6, 142 n.39
Blackaby, B., 230 n.50
Blaikie, P.M., 143 n.31, 143 n.44
Blalock, A., 42 n.5
Blalock, H.M., 42 n.5
Blome, D.A., 63 n.3
Blommestein, H., 180 n.148

Bobek, H., 141 n.4
Bologh, R.W., 278 n.32
Born, M., 16 n.29
Boudon, R., 188, 201 n.21
Boulding, K., 41 n.8, 95, 171
Bowden, L.W., 126 n.14
Bowers, G.H., 65 n.35, 179 n.112
Boyle, M.J., 203 n.39
Brackett, C.A., 63 n.5
Brail, R.K., xxvii n.15
Braithwaite, R.B., 15 n.9, 15 n.11, 41 n.4
Braverman, H., 277 n.20
Briggs R., xxviii n.22, 65 n.40, 151, 174 n.45, 260, 276 n.9
Britt, S., 63 n.3, 65 n.38
Brittan, A., 227 n.10, 227 n.13, 228 n.23
Brodbeck, M., 14 n.1
Brookfield, H.C., 210
Brown, J., 180 n.127
Brown, L.A., xviii, xxvii n.8, xxviii n.18, xxviii n.23, 17 n.39, 44, 60, 61, 63 n.9, 64 n.15, 91 n.17, 140 n.2, 140 n.3, 142 n.22–5, 142 n.28–30, 143 n.31–2, 143 n.35, 143 n.38, 143 n.40–1, 188, 190, 202 n.23, 202 n.33–4, 204 n.54–5
Brown, M.A., 142 n.27–8, 142 n.30, 143 n.41, 144 n.48–9, 144 n.57
Bruner, J., 42 n.19, 109–11
Buckley, W., 16 n.27
Bunge, W., 15 n.16, 276 n.2
Bunting, T., 97, 121 n.21, 105, 112, 122 n.37, 146, 147, 173 n.15, 177 n.83

Burke, C.J., 54, 65 n.36
Burnett, K.P., 60, 66 n.56, 140 n.59, 152, 158, 161, 168, 174 n.32, 175 n.52, 176 n.68, 177 n.79, 177 n.90, 178 n.103, 178 n.105, 179 n.106, 179 n. 119, 180 n.135, 180 n.139, 185, 200 n.10, 204 n.56
Burton, I., 120 n.1, 121 n.21
Bush, R.R., 59, 60, 65 n.43, 66 n.49
Butler, E.W., 189, 201 n.20, 202 n.24
Buttimer, A., xxix n.36, 112, 229 n.35, 247, 254 n.17, 261, 267, 276 n.10, 277 n.19, 278 n.30, 278 n.38

Cadwallader, M., 151, 174 n.33, 176 n.66, 177 n.79, 177 n.84, 179 n.119, 180 n.138, 202 n.35, 294 n.53
Campbell, D.T., 41 n.7
Canter, D., 120 n.4
Carlstein, T., 142 n.19, 143 n.37, 252 n.4, 253 n.9
Carrol, T.W., 141 n.14
Carroll, J.D., 57
Cederlund, K., 253 n.9
Cesario, F.J., 170, 181 n.147
Chapin, F.S., xvii, xxvii n.15, 201 n.20
Charles River Associates, 176 n.67, 177 n.78, 177 n.86
Christaller, W., 5, 63 n.6, 69, 90 n.3, 119
Clark, W.A.V., 55, 63 n.6, 90 n.3, 119, 191, 192, 195, 196, 200 n.13, 201 n.14, 202 n.35, 203 n.37, 203 n.42, 203 n.44, 203 n.51, 204 n.52, 204 n.58, 204 n.60, 205 n.69–70
Cofer, C.N., 65 n.42
Coombs, C.H., 57, 66 n.51, 86, 87, 90 n.4
Coombs, G.H., 64 n.32
Cooper, C., 229 n.39
Cox, K.R., xxviii n.24, 17 n.39, 100, 101, 102, 104, 113, 142 n.23, 146, 162, 166, 171, 172 n.10, 179 n.111, 180 n.129, 200 n.1, 205 n.71
Craig, S., 142 n.30
Craik, K.H., 175 n.60
Crothers, E.J., 65 n.35
Cullen, I.G., 147, 161, 169, 173 n.18, 173 n.20, 179 n.109, 180 n.146, 205 n.65, 254 n.14, 261, 267, 276 n.10, 277 n.19, 278 n.30, 278 n.38
Curry, L., 4, 7, 8, 15 n.7, 15 n.19, 15 n.22, 16 n.21, 16 n.23, 86, 91 n.17, 119, 159, 178 n.97

Cyert, R.M., 15 n.13

Dacey, M.F., 4, 12, 15 n.8, 16 n.36, 40, 42 n.28, 63 n.3, 147, 173 n.19
David, H.A., 90 n.4
Davies, R., 172 n.4, 175 n.55
Davis, K., 16 n.28
Davis, R.L., 64 n.32
De La Blache, P., 247
Dear, M.J., 205 n.70
Demko, D., xxviii n.22
Devlin, A., 173 n.14
Dobson, R., 158, 176 n.62, 177 n.90
Domencich, T., 175 n.53, 176 n.67
Donaldson, B., 192, 203 n.42
Dornic, S., 228 n.25
Downs, R.M., xxviii n.27, 17 n.41, 42 n.25, 120 n.4–5, 97, 121 n.14, 108, 112, 122 n.34, 122 n.39, 144 n.54, 146, 149, 162, 171, 172 n.12, 173 n.26, 173 n.31, 174 n.41–2, 210, 228 n.26
Dunbar, F., 177 n.90
Duncan, J.S., 219
Durkheim, E., 226

Ek, T., 253 n.9
Ellerman, D., 158, 161, 177 n.92, 178 n.105
Entrikin, N., xxv, xxix n.41, 112, 219
Ergun, G., 154, 176 n.63
Estes, W.K., 54, 64 n.24, 65 n.36
Evans, D.M., 221

Feller, W., 16 n.37
Ferber, R., 65 n.38
Fishbein, M., 41 n.14, 143 n.46, 144 n.47
Flowerdew, R., 204 n.63
Fourt, L.A., 65 n.38
Frank, L., 65 n.38
Frey, A., xxvii n.13
Frey, W.H., 188, 191, 195, 196, 197, 200 n.13, 201 n.20, 202 n.23, 203 n.36, 204 n.58, 204 n.62
Friedman, M., 15 n.12

Galanter, E., 66 n.49, 113
Gale, S., xv, xxvii n.6, 15 n.21, 121 n.11, 141 n.9, 276 n.3
Garrison, W., 63 n.11
Garst, R.D., 129 n.27
Getis, A., 64 n.18
Gibson, E., 221, 228 n.29, 229 n.43
Giddens, A., 229 n.33

Ginsberg, R., xxviii n.30, 190, 201, n.21, 202 n.31, 227 n.12

Godkin, M., 220, 254 n.16

Godson, V., 205 n.65

Goldstein, S., 183, 187, 188, 191, 195, 196, 197, 200 n.3, 200 n.13, 201 n.16, 201 n.20, 202 n.23, 203 n.36, 204 n.58, 204 n.62

Golledge, R.G., xxviii n.17, xxviii n.22, 15 n.5, 17 n.39, 44, 60, 61, 63 n.3, 63 n.9, 65 n.46, 66 n.58, 90 n.3, 91 n.17, 95, 100, 101, 102, 104, 113, 119, 121 n.14, 140 n.1–2, 141 n.13, 146, 162, 163, 165, 166, 167, 171, 172 n.10–14, 173 n.25, 179 n.111, 179 n.117, 179 n.120, 180 n.125, 180 n.129, 180 n.132, 200 n.1, 200 n.4, 200 n.12, 203 n.48, 204 n.53, 204 n.55, 205 n.7, 227 n.14, 228 n.17, 277 n.14

Goodknow, J.J., 42 n.19

Gorman, R., 227 n.11, 230 n.54, 230 n.57, 230 n.59

Gough, A.J., 276 n.4

Gould, P.R., xxvii n.9, xxviii n.20, 17 n.41, 42 n.23, 44, 47, 61, 63 n.8, 64 n.18, 66 n.57, 68, 90 n.1, 95, 96, 100, 101, 104, 171, 203 n.38, 203 n.49, 204 n.53, 260, 276 n.8

Graham, E., 97, 108, 173 n.17

Gregory, D., 121 n.11, 122 n.33, 228 n.30, 229 n.33, 278 n.26

Gross, S.R., 142 n.25

Guelke, L., 97, 105, 112, 121 n.21, 122 n.37, 146, 147, 173 n.15, 173 n.20, 177 n.83, 277 n.21

Guilford, J.P., 89, 90 n.4

Gulliksen, H., 90 n.4

Guthrie, E.R., 48, 49, 64 n.22–3, 113

Guttman, L., 70, 71, 90 n.5

Habermas, J., 229 n.33

Hagerstrand, T., xvi, xxii–xxv, xxvi n.3, xxvii n.8, xxvii n.11, xxvii n.16, xxviii n.21, xxviii n.31, 7, 8, 11, 15 n.18, 15 n.22, 16 n.31, 140 n.3, 141 n.5, 141 n.7, 141 n.10, 142 n.18, 142 n.21, 128, 131, 138, 160, 178 n.100, 204 n.54, 234, 252 n.3–4, 253 n.6, 253 n.10, 262, 277 n.12

Haggett, P., xxvii n.13

Haines, G.H., 60, 66 n.56

Hanneman, G.J., 141 n.14

Hanson, N.R., 14 n.3

Hanson, S., 158, 159, 161, 168, 169, 177–8 n.92, 178 n.94, 178 n.103, 179 n.106, 180 n.135, 180 n.139, 180 n.142

Hanushek, E.A., 189, 195, 196, 202 n.26, 204 n.59

Hanzik, C.H., 64 n.21

Harris, R.C., 228 n.29

Harris, R.S., 198, 205 n.68

Harrison, F., 144 n.56

Harrison, J.D., 174 n.42, 180 n.143

Harvey, D., xiii, xv, xxiv, xxvi n.2, xxvii n.5, xxix n.39, 15 n.5, 15 n.21, 16 n.35, 16 n.37, 41 n.2, 41 n.11, 119, 141 n.6, 166, 172 n.3, 179 n.111, 180 n.128, 180 n.130, 200 n.7, 212, 227 n.5, 228 n.17–18, 276 n.1

Harvey, M.E., 143 n.31

Hauser, J., 150, 174 n.39

Heggie, I.G., 158, 159, 160, 161, 178 n.92, 178 n.102, 179 n.108, 179 n.110

Hemmens, G.C., 201 n.20

Hempel, C.G., 15 n.24, 16 n.24, 91 n.19

Henley, D.A., 144 n.52, 204 n.57

Hensher, D., 158, 160, 161, 177 n.90, 178 n.92, 178 n.101, 179 n.107

Hensley, D., 175 n.61

Herskovits, M.J., 41 n.7

Hertzog, T., 179 n.121

Hightower, H.C., xxvii n.15

Hilgard, E., 63 n.2, 64 n.20

Himmelfarb, S., 143 n.46

Hollis, M., 228 n.18

Honikman, B., 144 n.55

Horkheimer, M., 252 n.1

Horton, F., 172 n.7

Houthakker, H.S., 88, 92 n.20

Howard, W., 180 n.143

Hudson, J.C., 6, 15 n.15, 141 n.15

Hudson, R., 120 n.4, 174 n.37

Huff, D., 44, 52, 63 n.10, 173 n.30

Huff, J.O., 191, 195, 196, 203 n.37, 203 n.51, 204 n.58, 204 n.60

Hull, C., 49, 51, 57, 64 n.14

Husserl, E., 112, 211, 224

Ide, E., 173 n.30

Isard, W., 40, 42 n.28, 63 n.3, 172 n.8

Jennings, D., 153, 175 n.58

Johnson, M., 158, 177 n.90

Johnston, R.J., 173 n.16, 175 n.56, 192, 203 n.42

Johnston, R.W., 277 n.13

Jones, P.M., 158, 160, 178 n.92, 178 n.102, 178 n.104

Kaiser, E.J., 201 n.20
Kalish, D., 64 n.18
Kaplan, A., 42 n.27, 101–2
Kaplan, R., 179 n.121
Kaplan, S., 122 n.41, 165, 179 n.121, 180 n.124
Kariya, P., 222
Kates, R.W., xxvii n.10, 42 n.29
Katona, G., 61, 64 n.13
Keat, R., 228 n.21, 229 n.32, 230 n.55, 230 n.60
Kelly, P., 138 n.54
Kendall, M.G., 91 n.14
Keppel, G., 65 n.42
King, L.J., 172 n.3, 200 n.12
Kirk, W., 41 n.9, 212, 276 n.5
Knapper, C., 29, 30, 42 n.12, 42 n.17, 42 n.25
Kneale, W., 16 n.30
Koopmans, T.C., 92 n.25
Koppelman, F., 150, 174 n.39
Koren, H., 230 n.58
Kruskall, J.B., 57, 66 n.52, 81, 82, 84, 90 n.4
Kuehn, A.A., 65 n.38
Kuhn, T.S., 14 n.2

Lambert, J., 230 n.50
Lenntorp, B., 178 n.100, 253 n.13
Lentnek, B., 178 n.100
Leontief, W., 92 n.20
Lerman, S., 158, 177 n.90
Levin, I., 158, 177 n.90
Levine, M., 54, 65 n.37
Lewin, K., 46, 64 n.16–17, 211, 212
Ley, D.F., xxv, xxix n.42, 97, 122 n.33, 172 n.2, 210, 217, 228, n.28, 229 n.33, 229 n.35, 229 n.42, 229 n.46, 252 n.2, 267, 268, 275, 277 n.23, 278 n.27, 278 n.29, 278 n.31, 278 n.37
Lieber, S., 144 n.53, 149, 157, 173 n.28, 175 n.35, 176 n.66, 177 n.80, 177 n.87, 181 n.149, 204 n.56
Lin, N., 141 n.14
Lloyd, R., 153, 175 n.58
Longbrake, D.B., 202 n.34
Losch, A., 4, 5, 63 n.6, 119
Louviere, J., 144 n.51–2, 147, 154, 157, 158, 173 n.21, 173 n.28, 175

n.61, 176 n.68, 177 n.87, 177 n.90, 195, 203 n.50, 204 n.57
Lovelock, C., 177 n.90
Lowenthal, D., 41 n.8, 42 n.24, 120 n.1, 100, 104, 105, 219
Lowman, J., 222
Lowrey, R., 151, 174 n.45
Luce, R.D., 57, 63, 66 n.49, 66 n.51, 92 n.21, 155, 169, 170, 176 n.64, 176 n.70, 180 n.141
Luckmann, T., 223
Luijpen, W., 230 n.58
Lukermann, F., 15 n.5
Lynch, K., 38, 42 n.26, 95, 101, 165, 171, 173 n.14

McCarthy, P., 150, 174 n.40
McFadden, D., 175 n.53, 176 n.67
McGinnis, R., 183, 188, 190, 200 n.4, 201 n.15, 201 n.17, 201 n.19
Macki, W., 91 n.8
Maher, C.A., 193, 203 n.47
Malecki, E.J., 142 n.25, 142 n.28, 143 n.36
Marble, D.F., 16 n.34, 44, 63 n.11, 127 n.17, 141 n.17
March, J.G., 15 n.13
Marchand, B., 252 n.1
Margenau, H., 14 n.3
Marx, K., 222, 226, 269, 270, 271, 275, 278 n.32–3, 278 n.35–6
Masnick, G., 201 n.15
Maxson, G.E., 143 n.41
Mayfield, R.C., 143 n.31, 143 n.38, 143 n.42, 144 n.58
Mead, G.H., 211, 219, 223
Mercer, D., xxix n.36, 97, 210, 229 n.35, 277 n.12, 277 n.22
Merleau-Ponty, M., 112, 211, 213
Merton, R., 121 n.12
Meyburg, A.H., 176 n.76, 177 n.90
Meyer, J.W., 129 n.26, 129 n.29, 142 n.26, 142 n.29
Meyer, R., 175 n.61, 177 n.90, 204 n.57
Michelson, W., 147, 173 n.22
Miller, G., 113
Mills, C.W., 216, 226
Misra, R.P., 126 n.13, 141 n.13
Mitchell, R.N., 63 n.3
Monty, R., 179 n.115
Moore, E.G., xviii, xxviii n.18, 188, 190, 198, 200 n.13, 202 n.23, 202 n.33, 205 n.68–70

Moore, G., 96, 98 n.14, 121 n.14, 204 n.53
Moore, G.T., 146, 162, 165, 172 n.13, 173 n.14, 180 n.125
Morgan, B., 176 n.73
Morrill, R.L., 16 n.32, 16 n.34
Morris, C., 31, 42 n.16
Mosteller, F., 59, 60, 65 n.43
Musgrave, B.S., 65 n.42
Muth, R., 202 n.28
Myers, C., 183, 188, 200 n.4, 201 n.15, 201 n.17, 201 n.19

Nadel, L., 110 n.30, 115 n.40, 122 n.30, 122 n.40
Nagel, E., 41 n.3, 228 n.19
Neisser, U., 179 n.112
Nell, E., 228 n.18
Nicolaidis, P., 175 n.51
Niedercorn, J., 172 n.8
Nijkamp, P., 181 n.148
Norman, K.L., 144 n.51
Norman, M.F., 65 n.44
Northrop, F.S.C., 41 n.3
Nourse, H.O., 66 n.48
Nystuen, J.D., 16 n.34, 141 n.17

O'Connor, K., 173 n.27
Ogden, C.K., 31, 42 n.15
O'Keefe, J., 122 n.30, 122 n.40
Olander, L., 253 n.9
Olsson, G., xiii, xv, xx, xxvii n.6, xxviii n.28, 15 n.5, 15 n.21, 16 n.25, 16 n.38, 63 n.4, 121 n.11, 168, 180 n.136, 200 n.7, 229 n.33, 252 n.2, 263, 276 n.3, 277 n.18
Osgood, C.E., 42 n.17, 42 n.25
Ostresh, L., 175 n.61

Paassen, C. van, 254 n.17
Pacione, M., 151, 174 n.48
Pahl, R., 198, 205 n.66
Paivio, A., 165, 179 n.113, 179 n.122, 180 n.123
Paris, C., 230 n.50
Parkes, D., 160, 178 n.98–9, 179 n.121, 252 n.4
Patricios, N., 149, 150, 152, 174 n.36, 174 n.41, 175 n.49, 177 n.85, 179 n.119
Perlmuter, L., 179 n.115
Persson, C., 254 n.18
Peterson, G.L., 69, 90
Philliber, S.G., 142 n.24

Piaget, J., 101, 108, 115, 116
Piccolo, J., 203 n.50
Pilger, J., 183, 188, 200 n.4, 201 n.15, 201 n.17, 201 n.19
Pipkin, J., 173 n.23, 175 n.54
Pitts, F.R., 16 n.31, 16 n.34
Pocock, D., 120 n.4
Polanyi, M., 114
Porteous, D., 120 n.4
Postman, L., 65 n.42
Potter, R., 153, 172 n.4, 175 n.57
Powell, J., 97, 210, 229 n.35
Pred, A., xv, xxvii n.7, xxix n.35, xxix n.37, 4, 8, 15 n.6, 16 n.22, 44, 63 n.4, 64 n.12, 141 n.16, 146, 172 n.6, 200 n.1, 204 n.64, 252 n.5, 255 n.32
Pribram, K., 100, 113
Pryor, R.J., 202 n.32

Quigley, J.M., 189, 195, 196, 200 n.13, 202 n.25–7, 204 n.59

Reibstein, D., 177 n.90
Relph, E., 219, 229 n.35, 278 n.25
Restle, F., 54, 65 n.37
Reynolds, D.R., xxviii n.24
Richards, I.A., 31, 42 n.15
Richards, P., 105
Riddell, J.B., 143 n.31
Rieser, R., 97, 143 n.43, 276 n.6
Ritchie, B.F., 64 n.19
Robertson, E., 143 n.34
Robinson, M.E., 203 n.39
Robson, B.T., 141 n.6
Roche, M., 230 n.58
Rogers, E.M., 141 n.14, 143 n.31, 143 n.33
Rommetveit, R., 42 n.17
Rose, C., 229 n.31
Rosenberg, S., 180 n.144
Rossi, P., 183, 187, 200 n.2, 201 n.15
Rowles, G., 220
Rushton, G., xvi, xvii, xxvii n.12, 17 n.40, 55, 62, 63 n.3, 66 n.58, 90 n.3, 91 n.8, 95, 97, 112, 119, 146, 147, 162, 163, 170, 172 n.11, 173 n.24, 176 n.66, 177 n.81, 179 n.116–17, 180 n.131, 200 n.8, 203 n.43, 215, 276 n.7

Saarinen, T., xxvii n.10, 42 n.29, 95, 97, 112, 120 n.4
Sabagh, G., 189, 202 n.24

Samuels, M., 97, 122 n.33, 172 n.2, 229 n.33, 230 n.44, 252 n.2
Sarre, P., 144 n.56, 174 n.42
Sauer, C.O., 103–4, 141 n.4
Sayer, D., 278 n.32
Schiller, R., 175 n.55
Schneider, R., 143 n.31
Schuler, H., 149, 151, 159, 174 n.34, 174 n.47, 176 n.66, 177 n.80, 178 n.95, 179 n.119
Schutz, A., 211, 223, 224, 225, 226, 228 n.20, 228 n.30
Seamon, D., 219
Segall, M.H., 41 n.7, 42 n.20, 42 n.24
Semple, R.K., 142 n.25
Shannon, G.W., 141 n.14
Shapira, Z., 176 n.74
Shapiro, P., 203 n.51, 204 n.60
Shaw, R.P., 200 n.13
Shepard, R.N., 57, 66 n.52, 80, 89, 90 n.4, 91 n.11, 92 n.24
Sheppard, E., 157, 168, 175 n.53, 176 n.68–9, 177 n.89, 192, 203 n.45
Sheskin, I., 181 n.149
Shoemaker, F.F., 143 n.33
Short, J.R., 193, 198, 205 n.67
Shrestha, M.N., 142 n.25
Shubik, M., 65 n.48
Siegel, S., 53, 65 n.34, 66 n.49
Silverman, D., 223
Simmons, J.W., 200 n.13
Simon, H.A., 15 n.13–14, 25, 41 n.6, 61, 66 n.49, 180 n.133, 180 n.144, 200 n.9
Skinner, B.F., 54, 168, 180 n.134, 212
Smith, C., 177 n.90
Smith, T., 168, 169, 170, 172 n.8, 180 n.140, 181 n.147, 203 n.40, 203 n.51, 204 n.52
Sonnenfeld, J., 42 n.24
Sopher, D., 229 n.39
Speare, A., 188, 191, 195, 196, 197, 201 n.13, 201 n.20, 202 n.23, 203, n.36, 204 n.58, 204 n.62
Spence, K.W., 57, 65 n.36
Spencer, A., 150, 152, 157, 174 n.38, 174 n.43, 175 n.50, 177 n.88
Spilmerman, S., 190, 202 n.30, 203 n.46
Stanfield, J.D., 141 n.14
Stanislawski, D., 141 n.4
Stea, D., xxviii n.27, 121 n.14, 104, 146, 162, 172 n.12, 173 n.26
Steele, J., 176 n.73
Stegman, M.A., 201 n.20

Steiner, I., 163, 179 n.118
Stevens, B.H., 63 n.5
Stever, J., 176 n.71
Stopher, P.R., 154, 175 n.51, 176 n.63, 176 n.76–7, 177 n.90
Strauss, A., 171
Suci, C.J., 42 n.17
Suppes, P., 57, 66 n.52, 176 n.64
Svart, L., 168, 180 n.136
Swinburne, R.G., 16 n.38

Tannenbaum, P.H., 42 n.17
Taylor, P.J., 98, 277 n.13
Taylor, S., 153, 175 n.59
Tennant, R.J., 55, 63 n.3–4, 90 n.3
Thomas, C., 175 n.55
Thomas, E.N., 63 n.3, 91 n.8
Thompson, D.J., 52, 65 n.33
Thorndike, E.L., 64 n.27
Thorson, S., 176 n.71
Thrall, R.M., 64 n.32
Thrift, N., xvii, xxvii n.14, 160, 178 n.99, 252 n.4, 253 n.5
Thurstone, L.L., 90 n.4
Tischer, M., 176 n.62
Tobler, W.R., 7, 8, 15 n.17, 47, 64 n.18, 109, 201 n.14, 276 n.2
Tolman, E.C., xix, xxviii n.26, 64 n.19, 64 n.21, 109, 113
Torgerson, W.S., 42 n.21, 57, 65 n.51, 90 n.4
Tornquist, G., xxviii n.25, 249, 251, 254 n.18
Tranter, P., 160, 178 n.98, 178 n.121
Tuan, Y.F., 96, 97, 100, 104, 105, 169, 173 n.16, 180 n.145, 219, 229 n.35
Tukey, J.W., 91 n.13, 92 n.21
Tullock, G., 277 n.15
Turner, M.B., 14 n.1
Tversky, A., 92 n.21, 170, 176 n.71

Underwood, B.J., 65 n.42
Urry, J., 228 n.21, 229 n.32, 230 n.55, 230 n.60

Van der Hoorn, T., 158, 178 n.92
Veenendaal, W., 181 n.148
Venezia, I., 176 n.74

Wales, H.G., 65 n.38
Warr, P.B., 29, 30, 41 n.12, 42 n.17, 42 n.25
Watson, J.B., 211

Webber, M.J., 172 n.9
Weber, M., 222, 225, 226
Weinberg, D.H., 189, 200 n.13, 202
　n.25, 202 n.27, 202 n.29
Weiss, S.F., xxvii n.15, 119, 201 n.20
Wheeler, J.O., 178 n.93
White, R.R., 17 n.41, 42 n.29, 96, 203
　n.38
Whorf, B.L., 42 n.20
Wilde, O., 116
Williame, R., 230 n.57
Williamson, F., 140 n.2
Wilson, A.G., 16 n.26, 41 n.1, 119, 172
　n.1, 172 n.9
Wohlwill, J.F., 174 n.44
Woldenberg, M.J., 15 n.20

Wolpert, J., xviii, xxii–xxv, xxviii n.19,
　xxviii n.29–30, xxix n.38, 8, 16 n.22,
　41 n.10, 44, 52, 61, 63 n.7, 183, 188,
　190, 191, 200 n.4, 201 n.15, 201 n.17–
　18, 202 n.22, 204 n.61, 262, 277 n.12,
　277 n.16
Wood, D., 165, 180 n.126
Woodlock, J.W., 65 n.38
Wright, J.K., 104, 225
Wrigley, N., 176 n.75

Yapa, L.S., 143 n.31, 143 n.38, 143 n.42,
　144 n.58
Yuill, R.S., 51, 64 n.29, 141 n.17

Zannaras, G., 173 n.25

Subject index

action space, 46–7, 87, 88, 153
activity, analysis, xvii; bundles, 234,
  236, 240–2, 244; system, 198, 233, 235
adoption, package, 135; rent, 132
aggregation, 80
alienation, 240, 270–1
analogy, 107–8
anthropocentricity, 265
aspiration, level, 48
attitudes, 30; attitude theory, 135–6, 139
attributes, generalized, 150
avoidance-conditioning models, 53, 59–
  60
axiom of cumulative inertia, 190
axioms, 5, 6, 9, 11; spatial, 5, 7;
  behavioral, 5, 13

bargaining, xxii
behavior, consumer, 44, 71–3; habitual,
  44, 45; in space, xvi, xvii, 184; least
  effort, 44, 45, 50, 51, 56; overt, 52;
  problem solving, 44–5; rational, 53;
  satisficing, 25–8, 44; search, 44, 49,
  68; spatial, xvi, xvii, xix, 184;
  stereotyped, 44, 49, 50, 51–2; trial and
  error, 49, 49–50, 54, 57, 59
behavioral geography, 100, 102, 104,
  105, 119; in Britain, xvii, xviii; in
  North America, xvii, xviii; matrix, 8,
  44; oscillation, 45
behaviorism, 211, 212

cause and effect, 6, 9
central places, 4, 5, 8

choice, 52–3, 55, 57–78; *a posteriori*,
  155, 169; axiom, 57–8; theory, Luce,
  155, 169
cognitive, behavior, 7; processes, 47; *see*
  cognitive map and cognitive mapping
cognitive map, 146–7, 167, 169; as a
  metaphor, 107, 109–10, 111; genesis
  of, 115–16; physiological basis of, 115
cognitive mapping, criticism of, 96–7,
  104–5; definition of, 99–100; relation
  to psychology, 117
concept identification, 53, 54–6
conjoint, measurement, 88; scaling, 149
constraints, authority, 235; capability,
  235; coupling, 235; travel, 157–8, 160
contiguity, 48–9
Cornell mobility model, 188
corporeal action, 242–4
correspondence rules, 4, 10
cues, 54–5

decision making, 43; processes, 52–3
determinism, 257, 261, 275
development, 131–3
dialetic, creation–destruction, 244–6;
  external, 242–4; individual–societal,
  246–7; internal, 242–4; path
  divergence–convergence, 244–5
diffusion, 8, 11, 12, 43; hierachical effect
  in, 125; market and infrastructure
  model in, 128–31, 139; neighborhood
  effect, 125
dissimilarities, 73, 80
distance perception, 150; cognitive

transformations, 151–2; estimation techniques, 151
distributions, 12

economics, neo-classical, 258
empiricism, 97–8, 112
entropy, 10
environment, cognition of, 101, 185; images of, 249–51; perceived, 259–60
exchange value, 272–3
existentialism, 210
experience, 242–3, 248
extinction, 49

genre de vie, 247
geographic models, 9
gestalt psychology, 211, 212
gravity model, 10, 151

habit, 49–51; family, 51; strength of, 49
HATS (Household Activity Travel Simulator), 161
hermeneutics, 218, 224
Housing Allowance Demand Experiment, 197
humanism, 97, 98, 112, 145, 209–30

ideal type, 224, 225
idealism, 266, 267, 275
ideology, 157, 168, 274–5
imagery, 153
indifference surface, 83
information, field, 10, 11, 125–6, 139, 153; integration theory, 137–8; processing, 167; processing and mobility, 195–7
instrumentalism, xxi, xxii, xxiv
intentionality, 212, 213
intentions, 242–3
interactance-process models, 53, 57–9
interactionism, symbolic, 219, 223
interpretation, *see* understanding

knowledge, external form of, 111; internal form of, 114–15; psychology of, 98, 101, 102

laws, morphological, 12; of chance, 11; of nature, 11; process, 12
learning, 8; and spatial behavior: definition of, 43; incremental, 49; latent, 48; models, 53–60; place, 47–8; sign, 47–8; theory, 12, 14, 43–4

life cycle, 189, 242
linear operator models, 53, 56
location theory, xxi, xxiv, 3, 6, 7, 13, 119, 257–8; economic, 20, 39–40; stochastic, 21, 39–40
locational analysis, 257–8
locational types, 71, 72, 86
logistic curve, 125

maps, mental, xix, xx, 13, 48, 96–7, 103–4, 108, 191–2
market, segmentation of, 154
Markov, 44, 49, 53, 188, 190; chains, 44; models, 53; processes, 49
marxism, xxiv, xxv, xxvi, 193
memory, 166
metaphor, 99, 107–8, 109–10, 119
mode choice, 148
models, nature of, 108–9
multinomial logit model, 152, 155, 157, 160, 162, 171
multipurpose trips, 159, 168–9

nature, transformation of, 245
naturalism, xxi, xxv, 212, 213, 266–7, 275
networks, associative, 165–6

operationalism, 112–13

paired associate models, 53, 56
paired comparisons, 70–1
participant observation, 210, 217, 220
paths, 234–5, 249, 251; daily, 241–2, 247; life, 241–2, 247
perception, 29–31, 43; environmental, xvi
personal construct theory, 138–9
personality, 153
phenomenology, 145, 210, 211, 223, 226; and idealism, 225; and institutions, 222, 223; and power, 222, 223; and triviality, 223, 224
philosophy, and geography, 97, 112–13
place utility, 190–1
positivism, logical, xxiv, xxvi, 211, 212, 214, 216, 217, 218, 259, 264–5, 269, 274
possibilism, 275
postulates, 5; behavioral, 5, 163, 166, 168–71; cognitive, 163, 166, 171; spatial, 5, 8
pragmatism, 210, 219

preference, functions, 13; ordering, 154, 170; revealed, 156, 192; subjective, 68–9; surface, 83
projects, 236–41, 245–6, 248, 251
provisional try, 50, 52

radical geography, 133–5
random numbers, 11
random utility model, 152, 154,162
rank size, 10
rating scales, 149, 151, 153
rational man, 44; economic, 44, 184; spatial, 44, 46
rationality, bounded, 44; economic, 24–8, 44
reductionism, 113–14, 213
regression, 10
repertory grid, 151, 156, 160, 166
response matrix, 87, 88
reward expectancy, 48

scale values, 79–82
scaling, models, 89; multidimensional, 62, 73–84, 150–1, 164; similarities, 79–84
search behavior, 196–7
semantic differential, 149, 156, 166–7
semiotics, 31–8
sign processes, 34–6
signs, theory of, *see* semiotics
simulation, 126; Monte Carlo, 10, 11, 14
social engineering, 236
sociology of science, 98, 100, 103, 105
space, non-Euclidean, 7

space–time models, 158–60, 162
spatial, analysis, 257; choice, 68–70, 147, 155, 158, 194; form, 7, 9; habits, 8; learning, 87; patterns, 12–14; processes, 7, 9
spread effect, 49, 237
stimulus ranking, 58–9
stimulus sampling models, 53
stochastic transitivity, 84, 89
stress, 81
surface images, 169
surrogates, 88
systems, large-scale, 10; small-scale, 10

technology, 248
theory, 4; assumptions of, 5; definition of, 4; higher-level, 5; learning, 12, 14; truth status of, 4
time geography, xxiii, 233, 241; dialectics of, 241–6
time paths, 241–4; daily, 241–2; life, 241–2; scale, 249
transitivity, 84–5
transportation planning, 155
trips, shopping, 148–50, 152; work, 145, 146, 148, 159, 166
two-step communication model, 132

understanding, 214, 218, 220, 222
use value, 272–3

verstehen, 218, 224
Vienna Circle, 4

Printed and bound by CPI Group (UK) Ltd, Croydon, CR0 4YY

22/10/2024

01777605-0014